Systems Thinking for Geoengineering Policy

Even by the scientists most closely associated with it, geoengineering – the deliberate intervention in the climate at global scale to mitigate the effects of climate change – is perceived to be risky. For all its potential benefits, there are robust differences of opinion over the wisdom of such an intervention.

Systems Thinking for Geoengineering Policy is the first book to theorise geoengineering in terms of complex adaptive systems theory and to argue for the theoretical imperative of adaptive management as the default methodology for an effective low risk means of confronting the inescapable uncertainty and surprise that characterise potential climate futures. The book illustrates how a shift from the conventional Enlightenment paradigm of linear reductionist thinking, in favour of systems thinking, would promote policies that are robust against the widest range of plausible futures rather than optimal only for the most likely, and also unlock the policy paralysis caused by making long-term predictions of policy outcomes a prior condition for policy formulation. It also offers some systems driven reflections on a global governance network for geoengineering.

This book is a valuable resource for all those with an interest in climate change policy, geoengineering, and CAS theory, including academics, under- and postgraduate students, and policymakers.

Robert Chris is a Visiting Fellow at the Department of Geography, The Open University, UK.

"Robert Chris convincingly describes the limitations of reductionist thinking when facing the complex challenge of developing appropriate policy for geoengineering research. By charting a way for geoengineering research policy that appreciates the features of complex adaptive systems, Chris moves beyond critique and opens our imagination to alternative courses of action."

– *Stefan Schäfer, Institute for Advanced Sustainability Studies (IASS), Potsdam, Germany*

"This book brings a welcome dose of fresh and constructive thinking to the debate about managing geoengineering technologies with a welcome focus on governing early innovation to manage risk without stifling creativity."

– *David Keith, Gordon McKay Professor of Applied Physics and Professor of Public Policy, Harvard University, USA*

"Radical measures to combat climate change are almost certainly necessary. This book is timely. It identifies the need for empirical research if policy decisions are to be based on knowledge and experience rather than ignorance and conjecture. The proposed "bottom-up" research governance regime would work with the grain of long-term uncertainty and unpredictability, thereby greatly reducing climate risks for future generations."

– *Hugh Hunt, Department of Engineering and Fellow of Trinity College, Cambridge, UK*

"In this bold critique of the 'predict-and-control' paradigm of policy-making the world has hitherto been relying on to deal with climate change, Robert Chris breaks new ground by deploying complex adaptive systems theory to the excruciating dilemmas of geoengineering. The implications for climate policy – and policy making in general – are potentially vast."

– *Olaf Corry, Associate Professor, Department of Political Science, University of Copenhagen, Denmark*

"In *Systems Thinking for Geoengineering Policy*, Robert Chris proposes a pragmatic approach for researching geoengineering as part of a comprehensive strategy for dealing with climate change that contrasts with the present unproductive approach of delaying any consideration indefinitely while all uncertainties are reduced and a top-down international governance system is established."

– *Mike MacCracken, Chief Scientist for Climate Change Programs with the Climate Institute in Washington DC, USA*

The Earthscan Science in Society Series

The Earthscan Science in Society Series aims to publish new high-quality research, teaching, practical and policy- related books on topics that address the complex and vitally important interface between science and society.

Vaccine Anxieties:
Global Science, child health and society
Melissa Leach and James Fairhead

Democratizing Technology
Risk, responsibility and the Regulation of Chemicals
Anne Chapman

Genomics and Society
Legal, ethical and social dimensions
Edited by George Gaskell and Martin W. Bauer

A Web of Prevention
Biological weapons, life sciences and the governance of research
Edited by Brian Rappert and Caitrìona McLeish

Nanotechnology
Risk, ethics and law
Edited by Geoffrey Hunt and Michael Mehta

Unnatural Selection
The challenges of engineering tomorrow's people
Edited by Peter Healey and Steve Rayner

Debating Climate Change
Pathways through argument to agreement
Elizabeth L. Malone

Business Planning for Turbulent Times
New methods for applying scenarios
Edited by Rafael Ramírez, John W. Selsky and Kees van der Heijden

Influenza and Public Health
Learning from past pandemics
Tamara Giles-Vernick, Susan Craddock and Jennifer Gunn

Animals as Biotechnology
Ethics, sustainability and critical animal studies
Richard Twine

Uncertainty in Policy Making
Values and evidence in complex decisions
Michael Heazle

The Limits to Scarcity
Contesting the politics of allocation
Lyla Mehta

Rationality and Ritual
Participation and exclusion in nuclear decision making, 2nd Ed.
Brian Wynne

Integrating Science and Policy
Vulnerability and resilience in global environmental change
Edited by Roger E. Kasperson and Mimi Berberian

Dynamics of Disaster
Lessons on risk response and recovery
Rachel A. Dowty Beech and Barbara Allen

The Social Dynamics of Carbon Capture and Storage
Understanding CCS Representations, Governance and Innovation
Edited by Nils Markusson, Simon Shackley and Benjamin Evar

Science and Public Reason
Sheila Jasanoff

Marginalized Reproduction
Ethnicity, infertility and reproductive technologies
Edited by Lorraine Culley, Nicky Hudson and Floor van Rooij

Resolving Messy Policy Problems
Handling conflict in environmental, transport, health and ageing policy
Steven Ney

The Hartwell Approach to Climate Policy
Edited by Steve Rayner and Mark Caine

Reconstructing Sustainability Science
Knowledge and action for a sustainable future
Thaddeus R. Miller

Experiment Earth
Responsible innovation in geoengineering
Jack Stilgoe

Rethinking Greenland and the Arctic in the Era of Climate Change
New Northern Horizons
Frank Sejersen

International Science and Technology Education
Exploring Culture, Economy and Social Perceptions
Edited by Ortwin Renn, Nicole C. Karafyllis, Andreas Hohlt and Dorothea Taube

Policy Legitimacy, Science and Political Authority
Knowledge and action in liberal democracies
Edited by Michael Heazle and John Kane

Systems Thinking for Geoengineering Policy
How to reduce the threat of dangerous climate change by embracing uncertainty and failure
Robert Chris

Systems Thinking for Geoengineering Policy

How to reduce the threat of dangerous climate change by embracing uncertainty and failure

Robert Chris

First published 2016
by Routledge

2 Park Square, Milton Park, Abingdon, Oxfordshire OX14 4RN
52 Vanderbilt Avenue, New York, NY 10017

Routledge is an imprint of the Taylor & Francis Group, an informa business

First issued in paperback 2019

British Library Cataloguing in Publication Data
A catalogue record for this book is available from the British Library

Library of Congress Cataloging in Publication Data
Chris, Robert, author.
Systems thinking for geoengineering policy : how to reduce the threat of dangerous climate change by embracing uncertainty and failure /
Robert Chris.
pages cm. -- (Earthscan science in society)
Includes bibliographical references.
1. Environmental geotechnology. 2. Climate change mitigation.
3. Environmental policy. I. Title.
TD171.9.C446 2016
363.738'746--dc23
2015016530

ISBN: 978-1-138-84117-8 (hbk)
ISBN: 978-0-367-27123-7 (pbk)

Typeset in Goudy
by Taylor & Francis Books

In memory of my father

Contents

Illustrations

Figures

Table

Acknowledgements

This book started life as a PhD thesis and it was Steve Rayner, editor of this series, who suggested I turn it into a book for publication. At best, only two people ever read a PhD thesis from cover to cover and they're the examiners who are paid to do so. Those of you reading this book will have different motivations. The purpose of a PhD thesis is to show how clever you are whereas the purpose of a book is to communicate, to engage others with your ideas. I hope that the radical surgery that the thesis has undergone to produce this book adequately respects the very different needs and interests of its new readership.

I must largely repeat the acknowledgements from the original thesis, as without their extraordinary support, none of this would have been possible.

My intellectual environmental journey began with the new millennium with James Lovelock's *Revenge of Gaia* and John Gray's *Straw Dogs*, and was consolidated by The Open University's undergraduate module U316, *The environmental web*. Little did I know then that a few years later some of its authors would become my PhD supervisors. At University College London Dr Ben Page and Dr Sam Randalls were instrumental in building my academic competence through the Environment, Science & Society Masters. At The Open University I must first thank all those involved in the postgraduate admissions process who had the faith to support my application through initial rejection to acceptance. Principal amongst these was Dr Jessica Budds whose encouragement and advice throughout that process were indispensible.

But my greatest academic debt is to my supervisors, Prof. Nigel Clark, Dr Dave Humphreys, and on Nigel's promotion to a Chair at Lancaster University, Dr Joe Smith. Their guidance, their insight, their encouragement, patience and extraordinary good humour, despite many provocations from my all too frequent failure to grasp, and even then heed, their advice, made my research an exhilarating, if at times painful journey into a promised land. Whatever academic credentials I might have now acquired, they are entirely due to their wise counsel and support.

I must also thank Dr Olaf Corry not only for his helpful comments, but also for sponsoring my appointment as a Visiting Fellow at the Open University without whose resources the further research necessary to produce this book would have been immeasurably more difficult.

Thanks are also due to Nigel Moore, Doug MacMartin, Rob Bellamy and most especially to Mike MacCracken, for their comments. I must also thank those at the Universities of Calgary, Heidelberg and Oxford for organising the geoengineering summer schools that I attended in Banff, Canada and Oxford, England. These were foundational experiences and many of the ideas presented in this book began their life in discussions at those fertile gatherings.

Notwithstanding all the advice, suggestions, and corrections, from those mentioned above and others too numerous to list, responsibility for the text rests with the author.

Finally, I must thank my wife, Val, whose enduring love, tolerance and forbearance have provided the space and time to allow me to indulge my academic passion.

Robert Chris
Tunbridge Wells, UK
May 2015

Notation

Within the literature there is great variety in the notation and language used to refer to the same concepts. Wherever possible I have endeavoured to use the standards listed below.

yr^{-1}	per annum
m^{-1}	per metre
m^{-2}	per square metre
m^{-3}	per cubic metre
ha^{-1}	per hectare (an area 100m by 100m)
~	approximately

Weights are generally presented in grams raised to a power of 10 by the standard SI unit.

Note that 10^{15}g = 1Pg = 1 Petagram = 1 Gigaton = 1Gt = 10^9 tonnes.

The following are the SI prefixes for large numbers:

k(ilo)	10^3 or 1 thousand
M(ega)	10^6 or 1 million
G(iga)	10^9 or 1 billion
T(era)	10^{12} or 1 trillion
P(eta)	10^{15} or 1 quadrillion
E(xa)	10^{18} or 1 quintillion
Z(etta)	10^{21} or 1 sextillion
Y(otta)	10^{24} or 1 septillion

Mtoe (Figure 1.5) means millions of tonnes of oil equivalent. 1Mtoe is the amount of energy released by burning one million tonnes of crude oil.

Btu (Figure 9.1) refers to the British Thermal Unit and is also a measure of energy. 1Btu is the amount of energy needed to cool or heat one pound of water by one degree Fahrenheit.

Mtoe and Btu have largely been superseded by kilowatt hours (kWh) in academic literature but are still used by some in the energy sector. 1MBtu = 293kWh and 1Mtoe = 11,630GWh.

Acronyms

AOGCM	Atmosphere Ocean and General Circulation Models
AOSIS	Alliance of Small Island States
BAU	Business as usual
BECCS	Bio-energy with carbon capture and storage
BPC	Bipartisan Policy Center
C	Carbon
C_{eq}	Carbon equivalent (a means of aggregating carbon and non-carbon GHGs)
CAS	Complex adaptive system
CCS	Carbon capture and storage
CDIAC	Carbon Dioxide Information Analysis Center
CDR	Carbon dioxide removal
CO_2	Carbon dioxide
COP	Conference of the Parties, UNFCCC
DACS	Direct air capture and sequestration
DICE	Dynamic integrated climate-economy
FAO	UN Food and Agriculture Organization
FAQ	Frequently asked questions
GGN	Geoengineering Governance Network
GHG	Greenhouse gas
GMST	Global mean surface temperature
IPCC	Intergovernmental Panel on Climate Change
LERC	Cardiff University's Lean Expertise Research Centre
LULCC	Land use and land cover changes
LWR	Long wave radiation
NAS	US National Academy of Sciences
NASA	National Aeronautics and Space Administration
NERC	Natural Environment Research Council
NPP	Net primary production
PAGE	Policy Analysis of the Greenhouse Effect
ppmv	Parts per million by volume
PDF	Probability distribution function
RICE	Regional integrated model of climate and the economy

SAI	Stratospheric aerosol injection
SPM	Summary for Policymakers
SRM	Solar radiation management
SWR	Short wave radiation
TCS	Theory Culture & Society
TEU	Twenty foot unit (38.5m^3)
UNFCCC	United Nations Framework Convention on Climate Change
WGIII	IPCC Working Group III (Mitigation of Climate Change)
Wm^{-2}	Watts per square meter
WWII	Second World War

1 Contextualising geoengineering

> The paradox is this – humans are the only animal of which we are aware that is capable of cumulatively enhancing its knowledge and its power, but they are also an animal that is chronically incapable of learning from its experience.
>
> John Gray[1]

Publication of this book coincides with COP21, the UNFCCC gathering in Paris in late 2015 of more than 40,000 of the global environmental elite, for the largest international conference France has ever hosted. The UNFCCC's objective for COP21, after a quarter of a century of sustained international effort, is finally:

> to reach, for the first time, a universal, legally binding agreement that will enable us to combat climate change effectively and boost the transition towards resilient, low-carbon societies and economies. To achieve this, the future agreement must focus equally on mitigation – that is, efforts to reduce greenhouse gas emissions in order to limit global warming to below 2°C – and societies' adaptation to existing climate changes.[2]

This book makes two interrelated arguments. First, there is now sufficient credible evidence to suggest that if this objective is not already unattainable, it is, at best, extremely challenging and its realisation is increasingly likely to require some assistance from geoengineering, a range of technologies aimed at reducing the effects of global warming by direct intervention in the global climate. Second, the framing of climate change as a problem that is amenable to solution by a legally binding agreement seriously undermines attempts to reduce its risks because this framing is structurally incapable of accounting adequately for the dynamic complexity of the ecosphere of which humanity and the climate are only part. Drawing these two arguments together, the book proposes a paradigm shift, transcending the conventional linear and reductionist problem solving mindset that has characterised the scientific method since the onset of the Enlightenment, in favour of a systems thinking approach that embraces the inescapable uncertainties inherent in natural systems and respects the autonomy of the future generations whose interests lie at the heart of our concerns about climate change.

In the epigraph above, Gray was not suggesting that humans never learn from experience but rather that our learning has been erratic and discontinuous. All too often, he implies, we trample on or ignore giants of the past rather than stand on their shoulders. While recognising the exponentially accumulating body of human knowledge and extraordinary technological advances, Gray asks whether the same applies to the quality of human behaviour. Even allowing for some intergenerational variability, he questions whether humanity is becoming progressively more compassionate and wise, and whether our behaviour is increasingly driven by genuine insight rather than folly. We can apply Gray's question about the quality of human behaviour to climate change and also ask whether attempts to intervene directly to control the global climate would amount to a folly, or perhaps the converse, that now, because of the delay in addressing the problem effectively, not to do so would be a folly.

In *The March of Folly: From Troy to Vietnam* Barbara Tuchman (1984) defines a folly in public policy as one the pursuit of which is 'contrary to the self-interest of the constituency or state involved'. To qualify as folly a policy must have four specific attributes. First, it must have been perceived as counterproductive in its own time, not merely in hindsight; second, a feasible alternative course of action must have been available; third, it must be that of a group, not an individual ruler; and finally, it should persist beyond any one political lifetime (1984, 5). Mapping these conditions onto climate change it is clear that the continued emission of large amounts of anthropogenic greenhouse gases (GHGs) already qualifies as a folly. What makes this particular folly so intractable is that the 'group' is so broad, encompassing not just a particular political elite or dynasty, but the political elites of all the developed economies over the last several decades, and also their counterparts in the even larger developing economies that aspire radically to improve the lifestyles of their constituents. This folly is systemic.

Since the 1970s there has been accumulating scientific evidence that human activity is increasingly responsible for changing the global climate. As the science has become better understood, we have come to realise that these changes are happening at an unprecedented rate. However, the numbers of people whose lives have been seriously affected is still relatively small and the effects on other life-forms go largely unnoticed by the majority of the global population whose urban lives distance them from the realities of the natural world. Yet, as a result of an extraordinary global scientific endeavour spanning the last three decades there is now sufficient evidence to know beyond reasonable doubt that it would be a folly of epic proportions to ignore the threat of unabated climate change by not reducing our carbon footprint. The open question is the extent to which our response to climate change is to be driven by compassion, insight and wisdom rather than their opposites.

We have grown accustomed to coping with the fickleness of the weather. From drought to flood, from heat wave to mid-summer snowfall, the weather has always been a key determinant of famine and feast, of joy and misery. But this has not been a one way street. The relationship between humans and the weather is symmetric, both reflexively influencing each other. Until recently, human influence on the weather was inadvertent, a side effect, an unintentional consequence of human activity. For tens of thousands of years this was largely due to land use changes with, for example, vast areas of forest being cleared and hillsides terraced for agriculture. But more recently, it has been due to industrialisation and urbanisation. Now, just 300 years after the invention of the steam engine, fear of the consequences of escalating anthropogenic climate change is opening up the prospect of recasting human influence on the weather from being unwitting to being deliberate.[3] In keeping with the prevailing techno-Zeitgeist, the quest for technological solutions to avert the threat of dangerous climate change is now well established. This has provoked investment across the economy, including in low-carbon energy technologies, new energy conservation standards in construction, and changes in agricultural and other land use practices. It has also spawned a diverse range of technologies designed to intervene directly in the global climate system in order to mitigate the effects of global warming. These technologies are known collectively as *geoengineering*.[4]

Something new under the sun

Climate change is important because all life depends upon a degree of stability in its habitat and that stability is undermined by rapid climate change. It is for this reason that the UNFCCC has enshrined in its Article 2 the objective of:

> stabilization of greenhouse gas concentrations in the atmosphere at a level that would prevent dangerous anthropogenic interference with the climate system. Such a level should be achieved within a time frame sufficient to allow ecosystems to adapt naturally to climate change, to ensure that food production is not threatened and to enable economic development to proceed in a sustainable manner.[5]

Faced with the threats contemplated in this Article, there are broadly three responses: a) *mitigation* or *emissions abatement* – reduce the behaviours that create these conditions at sufficient scale and speed to offset the threat;[6] b) *adaptation* – adapt to the changes and/or cope with the threats as they crystallise; and c) *geoengineering* – intervene in the climate system to avert or reduce the threat.

These options are not mutually exclusive. In the early stages of the UNFCCC the focus was entirely on emissions abatement. The logic followed classic reductionist thinking – if the problem of climate change is caused by anthropogenic greenhouse gas (GHG) emissions, then the solution must be to reduce those emissions. It then became apparent that that was not such an easy thing to do at the scale and speed necessary to avert the feared threats, so the idea of adaptation, that had previously been regarded as something of a distraction, was promoted as an additional response. More recently, concerns have arisen that the combined effects of emissions abatement and adaptation might still not be enough and that some geoengineering might be necessary.

It is often said that there's nothing new under the sun, some of us are just obsessed with bigger, better, faster, cheaper ways of doing what has always been done – cultivating, making, consuming, trading, dressing, travelling, communicating, fighting, loving, hating and generally engaging in a range of behaviours that have characterised the human condition since before we were human. And we are constantly producing novel ways of doing these things – cars have replaced horses as a mode of transport; mobile phones have replaced smoke signals as a means of long distance communication; guided missiles have replaced catapults as a means of warfare, the list is endless.[7] But how often do we come up with a genuinely new thing to do? Deliberately engineering the global climate is, I contend, a genuinely new thing to do. There is no other activity in which humanity has proposed a deliberate intervention in planetary scale processes that would directly, and more or less simultaneously, affect almost every living creature. All other interventions in nature have been local and intended to serve some conventional purpose – food, shelter, security and so on. Engineering the climate is not a new way of doing something we've always done; it's an entirely new idea – to turn the entire climate system into an artefact in order (hopefully) to render it benign.

The rhetoric

This small selection of quotes spanning almost thirty years illustrates that the political rhetoric has been directed at framing our response to climate change first and foremost as an obligation to future generations.

> *Sustainable development satisfies the needs of the present generation without compromising the chance for future generations to satisfy theirs.*
>
> Brundtland Report 1987

> *We are … the trustees of this planet, charged today with preserving life itself – preserving life with all its mystery and all its wonder.*
>
> Margaret Thatcher, 1989 speech to UN General Assembly

> *The Parties should protect the climate system for the benefit of present and future generations of humankind.*
>
> UNFCCC 1994 Art. 3.1

> [The] *modelling framework has to take into account ethical judgements on … how to treat future generations.*
>
> Stern Review 2006:ix

Les générations futures ne nous le pardonneraient pas [*l'échec de l'accord mondial sur le réchauffement climatique*].

French President Sarkozy, interview in *Svenska Dagbladet* 3 July 2009[8]

Der Kampf gegen die Erderwärmung ist eine 'Überlebensfrage der Menschheit'.

German Chancellor Angela Merkel[9]

We can lead the world, secure our nation[*'s energy needs*], *and meet our moral obligations to future generations.*

Candidate Barack Obama, 2008 election campaign speech on energy policy

it's all about passing on things to future generations, holding the planet in trust.

David Cameron, 2008 interview in *The Observer*

Now is the moment when our responsibility to future generations must be answered.

Tony Blair, 2009 Speech to the MASDAR World Future Energy Summit

to think in terms of epochs and eras and how our stewardship [*of the planet*] *will be judged not by tomorrow's newspapers but by tomorrow's children.*

Gordon Brown, 26 June 2009 launching The Road to Copenhagen

for the sake of our children and our future, we must do more to combat climate change.

President Barack Obama, State of the Nation address 12 February 2013

The message is clear. The rhetoric is that future generations are part of our moral community and we owe it to them to prevent climate change from undermining their enjoyment of the benign future we wish for them.

There are a number of significant questions embedded within this aspiration. Who are *we*? Who are *they*? What does this benign future look like, and how are we to know the extent to which our actions make more likely its coming to pass without simultaneously creating other intractable problems in its wake? How do we assess and then rank the cost of any sacrifices we must make to secure this benign future, against the benefits they might receive from it? Are there any no-regrets policies that could deliver this benign future to more distant future generations without diminishing the well-being of the present and more proximate future generations? These would correspond to what economists refer to as Pareto optimal outcomes in which everyone is better off and no one worse off. Given that all prognostications about the future are subject to increasing uncertainty the further into the future they refer, how are we to devise appropriate near-term policy responses? These are just a few of the many challenging ethical and practical questions emerging from the political rhetoric. These are not questions that have definitive answers.

When I started this project my focus was on the ethical dimension of our relations with the future generations who were the putative primary beneficiaries of our awakening consciousness of the dangers from climate change. Important, indeed vital, as the ethics is, I gradually came to the view that while policymakers might fine tune a policy to reflect ethical concerns, such concerns are rarely the primary drivers of policy. Ending judicial capital punishment, legalising abortion, or the adoption of gay marriage, these are public policies driven by and decided on largely ethical grounds, but pervasive changes to the global political economy, an inevitable consequence of responding to climate change at scale, are multi-faceted complex propositions in which ethics is but one of many considerations, and possibly not the most significant.

Key world leaders have acknowledged the subordination of climate ethics to the primacy of the markets and economic factors. In the paragraph before Obama's moving plea (cited above) on behalf of 'our children and our future', he stresses how responding to climate change has already produced 'tens of thousands of good, American jobs' and in the following paragraph he urges Congress to pursue a 'market-based solution to climate change'. Similarly, in a 1990 speech to the UK Royal Society shortly after her ground-breaking appeal to the UN General Assembly (also cited above), Margaret Thatcher qualified her concerns for the 'mysteries and wonders' of life, stressing that:

> to deal with environmental problems, we must enable our economies to grow and develop, because without growth you cannot generate the wealth required to pay for the protection of the environment.
>
> (later quoted by herself, Thatcher 2002, 452)

She then assured the Royal Society that: 'We can rely on industry to show the inventiveness that is crucial to finding solutions to our environmental problems'. She did not specify precisely what the market drivers were that would redirect industry's inventiveness from creating environmental problems to resolving them. Nor did she define 'our environmental problems' in precise terms, nor specify how we, or our successors, would know that they had been solved. In her memoir *Statecraft* she refers to the anti-capitalism and anti-Americanism 'which always lay under the surface of environmentalism' (ibid.).

It seems clear that however much we care about future generations and the 'mystery and wonder' of life, climate change and responses to it are more than merely a matter of the physics and chemistry of the weather; the needs of future generations are to be moderated by the market. However, if that is so, perhaps we should first establish whether the market might be part of the problem rather than part of the solution, particularly if, as some believe, 'infinite growth is not possible on a finite planet' (Moore 2015). If this is so, there is some conflict between national leaders' quest for limitless growth and the demands of averting dangerous climate change that requires thinking outside the prevailing neoliberal paradigm. But this is scary precisely because a new paradigm introduces new risks and uncertainties that take us out of our comfort zone.

The principal effort internationally in combating climate change continues to be the UNFCCC process that began in the early 1990s.[10] The graph in Figure 1.1 illustrates the lack of success it has had in stemming the escalating growth in CO_2 emissions. In the 262 years from 1750, the year generally taken as the beginning of the industrial age, global anthropogenic fossil fuel CO_2 emissions totalled in excess of 380PgC[11] and a further 170PgC[12] has been emitted as a result of land use and changes in land cover. Of the fossil fuel and cement production emissions, half was emitted in the last thirty years. The scale of the task of controlling emissions is further illustrated in Figure 1.2 by extrapolating this historical trend. The Business As Usual (BAU) scenario would result in annual emissions exceeding 45PgC by 2100, almost five times the current level; cumulative emissions would also be five times their current value and more than two and a half times the 1,000PgC considered to be compatible with limiting to 2°C, the global mean surface temperature (GMST) increase since the beginning of the industrial era (IPCC 2014, sec. 12.5.4). In this scenario, the last forty years would account for half of the emissions for the entire period of 350 years from 1750 and the 1,000PgC cumulative emissions limit would be breached in 2043. Of course, there are

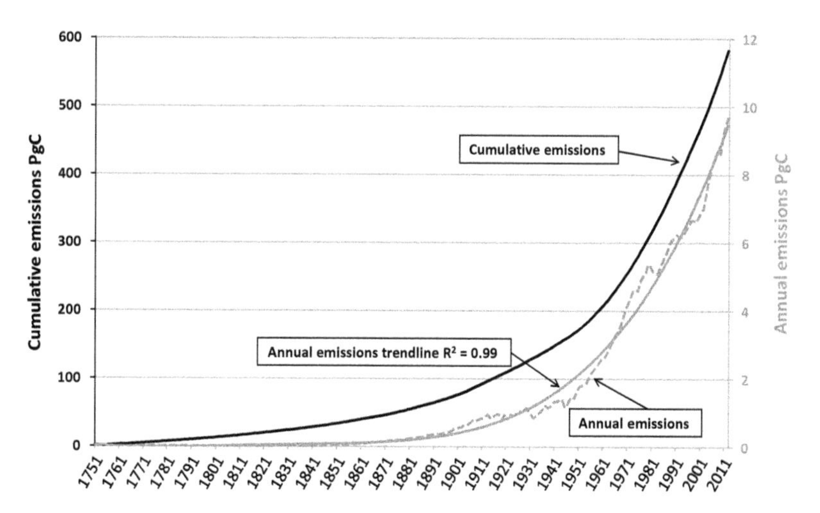

Figure 1.1 Global fossil fuel CO_2 emissions 1750–2012.
Source: data from CDIAC

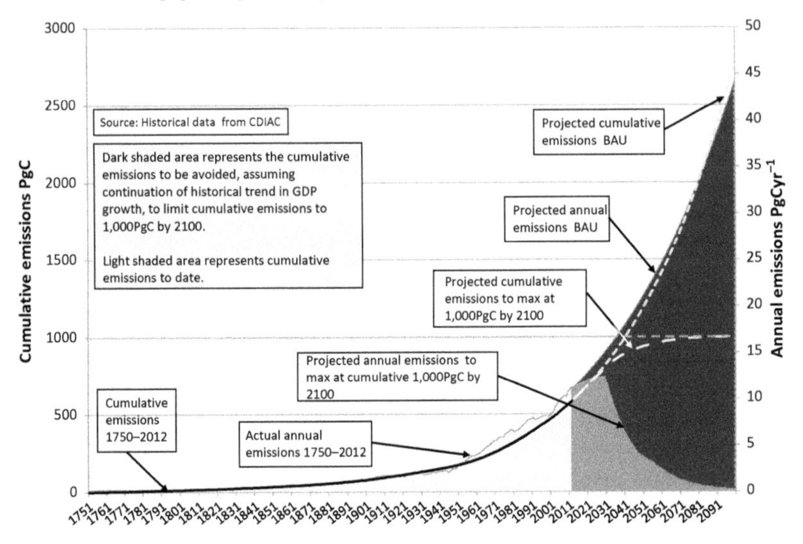

Figure 1.2 Total CO_2 emissions 1750–2012 and projections to 2100
Source: historical data from CDIAC; projections extrapolated by author from a fourth order polynomial regression with R^2=0.99. Emissions from land use change prior to 1850 are not available and have been ignored as being insignificant; after 2005 these emissions are assumed to continue at their current rate of ~1.5PgCyr^{-1} until 2100.

many reasons why emissions over the next eighty five years might not follow the same trajectory as the past, but these figures indicate the scale of the change in human behaviour that is implied in the emerging discourse on averting dangerous climate change.

In this model, the emissions reduction to bring cumulative emissions to 1,000PgC by 2100 and emissions to zero by then, assume that emissions will grow increasingly more slowly in the coming years, peaking in 2030. Thereafter, they fall rapidly but at no more than 6 per cent each year. On these projections, cumulative emissions will have reached 773PgC by 2030, leaving only 227PgC for the remaining seventy years, an average of little more than 3PgCyr^{-1} supporting GDP up to five times its current level, against current emissions of ~11PgCyr^{-1}, implying a more than eighteen-fold improvement in carbon intensity. Because these figures include emissions of ~1.5PgCyr^{-1} from land use change, which is assumed to continue indefinitely, stabilisation at 1,000PgC implies routine CDR geoengineering continuously to remove this amount of atmospheric carbon.

How apocalyptic breaching the 1,000PgC limit would be is open to debate as is the time it would take for this supposed apocalypse to materialise, but it must be remembered that the gates of climate hell do not suddenly open as we pass any particular milestone. The inertia in the climate system is such that it may be many decades before the worst effects of global warming become apparent. But we must also remember that the climate system is chaotic, and may surprise us with one or more abrupt changes along the way.

Climate intervention

It has been in response to the magnitude of the emissions abatement task that there has been increasing interest in adaptation and geoengineering. The novel idea of humanity consciously taking control of the global climate system introduces some novel questions. Some of these questions are ethical and address concerns about whether, as a matter of principle, we should usurp nature's power in this way, and if it were decided that we should, what moral criteria might apply in determining how it should be done. Other questions focus on the practicalities. For example, climate scientists are busy trying to establish what interventions would deliver what changes to the climate. Other scientists ask what the effects of those changes would be on, for example, biodiversity, the economy, and human well-being. Engineers consider how the machines to perform those interventions might be made to be climatically effective yet safe. Responding to all these questions demands knowledge and understanding of natural, technological and social systems and their interactions. Without prejudging the outcome, this book is first and foremost about the epistemological challenges facing those engaged in turning the concept of geoengineering into reality. How do they acquire data and knowledge to inform these interventions? Once policymakers had crossed the political Rubicon and decided to intervene in the global climate system, how would they determine how best to do so? And could a global consensus emerge that would produce a unified global geoengineering policy, and if not, would that be a bar to action?

Geoengineering is an entirely new, and as yet untried, endeavour. There is no empirical record for policymakers to call upon. If for all its potential benefits, geoengineering also entails some significant risks, how are policymakers to realise the former without increasing exposure to the latter? There are alternative ways to determine what constitutes the most attractive policy option. It could be that which best corresponds with the facts or reality, that which coheres best with generally held beliefs and values, or that which, in the long run, delivers the best outcome. These three categories correspond to Durand *et al.*'s (2006) distinction between the *correspondence*, *coherence* and *instrumentalist* theories of truth. The correspondence theory is associated with positivist and realist approaches to epistemology. For geoengineering, such an approach is undermined by the difficulty of establishing 'facts' and 'realities' associated with distant future states of nature. For coherence theory, the time scales involved in climate change make it problematic to establish any consensus about beliefs and values that are held by actors separated by multiple generations, each of which is characterised by a multiplicity of cultural value sets.

Instrumental truth is a pragmatic epistemological approach but is also subject to similar limitations. Given the extended time frames of climate change and geoengineering, the great likelihood is that any benefits will not be enjoyed by those who initiate the policy but by their distant descendants.

This introduces two considerable problems. First, it cannot be assumed that the policy initiators' beliefs about what would constitute a benefit would coincide with those of the distant future supposed beneficiaries of those policies. Second, when it comes to assessing causal chains linking near-term policy with long-term effects, distinguishing between the merely plausible and the likely becomes increasingly difficult as the temporal and spatial horizons expand. If the pragmatist approach to epistemology is to overcome these difficulties, it must find ways of making truth statements about the distant future outcome of near-term policy options. It will become clear that it is precisely this difficulty that is at the heart of the dominant framing of climate change policy, explaining why the international community has yet to fashion a timely and effective response to the anthropogenic forcing of the climate, and why, notwithstanding the increasing urgency, it remains an elusive and challenging goal. It also explains why the idea of direct climate intervention at a global scale remains a low priority for policymakers. If the pragmatist approach to truth is to prevail and result in effective policy outcomes for climate change, I shall argue that a paradigm shift to systems thinking is required.

Geoengineering

The graph in Figure 1.3 shows that the academic engagement in geoengineering grew rapidly following Nobel laureate Crutzen's 2006 article registering increasing concern at the continuing lack of effective action by policymakers

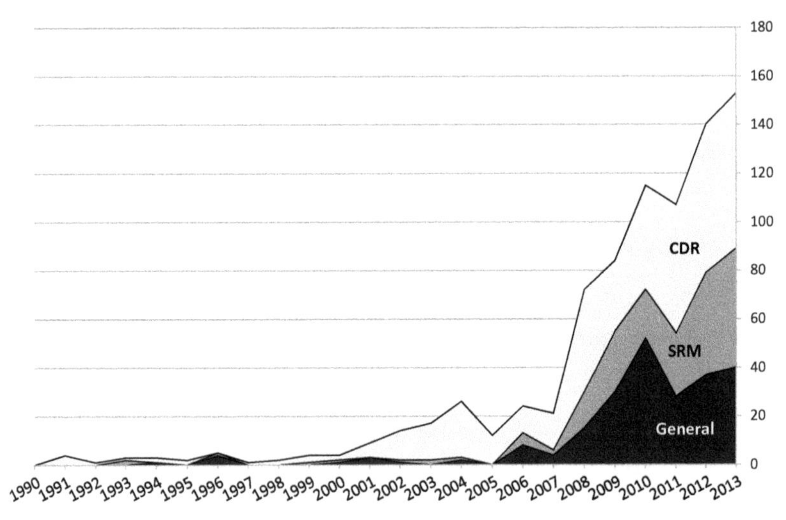

Figure 1.3 Publication trends by thematic types for SRM, CDR and 'general geoengineering'

Source: adapted from Oldham *et al.* 2014.

to limit GHG emissions. Notwithstanding these concerns, that have been more widely expressed since then, there is scant evidence of any serious interest by policymakers in geoengineering as an additional policy option.

Climate scientists likened geoengineering to the emergence of other novel technologies such as genetic engineering and nanotechnology, and argued that if there were a general perception that the governance was in place to control risks and ensure that proper ethical standards were observed in any research and development, policymakers and the public might be more receptive to supporting further research. Conferences were held in 2010 to explore these ideas in both the US and the UK with participation from many academic disciplines and the wider community including the media and environmental NGOs ('The Asilomar Conference Recommendations on Principles for Research into Climate Engineering Techniques' 2010). The House of Commons in the UK and the House of Representatives in the US undertook a joint enquiry into geoengineering; their reports are discussed in Chapter 6.

In order to understand better the risks entailed by geoengineering, the climate science community began to focus increasingly on predictions derived from climate models. The underlying argument seemed to be that if sufficient theoretical knowledge were available, or risk and uncertainty could be reduced to an acceptably low level,[13] policymakers might then see geoengineering as a more viable option. However, this theoretical research was usually, and from a scientific perspective rightly, accompanied by caveats about the possible unintended adverse consequences of geoengineering and the paramount importance of pursuing vigorous emissions abatement. This positioning not only emphasised the risk profile of geoengineering but also presented it as a lower priority than emissions abatement. It continues to this day with the NAS declaring that 'there is no substitute for dramatic reductions in CO_2 emissions to mitigate the negative consequences of climate change' and that certain types of geoengineering 'introduce novel risks' (NAS 2015). With this precautionary ambivalence from the science community, however intellectually justifiable, it is unsurprising that interest in policy circles in geoengineering was, and continues to be, more in the nature of a 'watching brief' than active engagement.[14]

There was also much talk about geoengineering presenting a so-called *moral hazard* that might encourage policymakers to adopt it as a cheap quick *techno-fix* despite its many feared unintended side effects, rather than pursue the more benign, but possibly more radical, costly, and therefore more politically challenging option of the rapid decarbonisation of the global economy (Bracmort, Lattanzio, and Barbour 2010, 7; Cicerone 2006, 224; Virgoe 2009, 107). But Bunzl argues cogently that:

> Moral hazard only arises for geoengineering if you think that research or, if it came to it, implementation would undermine other actions and lead to more not less greenhouse gas output. That seems far-fetched

> since, at least among policy makers, nobody believes that geoengineering offers anything but a relatively short stopgap to buy time for other action. Nor are the funds that would be needed for geoengineering research large enough relative to the research budgets of even the United States, let alone the whole developed world, to create an allocation issue.
> (Bunzl 2009, 2)

Notwithstanding Bunzl's view that geoengineering is not a threat to other actions to reduce GHG emissions, a political impasse for geoengineering was emerging. Why would policymakers support geoengineering, a novel strategy with complex international implications, when they were still committed to the long-running UNFCCC process? Despite its travails, no one at the UNFCCC has promoted geoengineering as a viable alternative or complementary response to climate change. A search for 'geoengineering' and 'climate engineering' on the COP21 website yields no results.

It is tempting to dismiss geoengineering, as many of its detractors have, as a step too far, a literal opening of Pandora's box unleashing all manner of environmentally driven evils, and heralding a future in which the climate becomes a violently contested political space (Marshy 2010). This book does not argue for the desirability of geoengineering but rather suggests that it would be prudent to accept the possibility of it being a useful adjunct in the climate policy portfolio mix, and proposes how this might be explored in a way that keeps the lid on the box. Nevertheless, a decision by those in the future to deploy geoengineering would be a recognition of a monumental failure of climate policy by earlier generations. The sheer idea of geoengineering is preposterous and little intellectual investment is needed to understand that the planet would be better off were it not needed. Unfortunately, the accumulating evidence suggests that rejecting it summarily might be unwise (NAS 2015).

Preposterous though it may be, geoengineering is generating increasing interest and with this has come a great deal of controversy. Ethicists argue about its morality particularly with regard to the interests of future generations (Gardiner 2011; Preston 2012), sociologists argue about its politics and governance (Barrett 2014; Szerszynski *et al.* 2013), and about the scientisation of policy (Stilgoe 2015). Some argue about its effectiveness (Bala 2009; Brovkin *et al.* 2009), while others argue about its underlying wisdom (Fleming 2010; Hamilton 2013; Robock 2008; Pierrehumbert 2015). Geoengineering has become a space in which academics and civil society activists of all stripes can engage their imaginations by exploring all manner of potential pitfalls – no empirical evidence is necessary to participate. There are rarely right answers to the many challenging questions that these debates address. Nevertheless, they raise important questions and it is also important that they be reflected upon because even if the climate engineers were able to deliver the means of intervening in the global climate to avert dangerous climate change, that would not be a sufficient reason to do it.

This book runs along a parallel strand. It does not engage directly with these questions but asks another set of questions concerned with policy formulation. My starting point is some indeterminate time in the future when the cumulative efforts to avert dangerous climate change look increasingly inadequate to the task. Maybe climate change will have progressed faster than we currently expect. Maybe responses to climate change will have proceeded more slowly than we currently hope. Maybe the responses will have proceeded as planned but will not have been as effective as anticipated. There are any number of reasons why those in the future might see the threats from climate change as more imminent, more severe, less avoidable. What will their attitude be to geoengineering? Might they be grateful that their predecessors had prepared the ground so that they would be armed with this additional policy option? Might they lament that their predecessors had failed to do sufficient either to avert dangerous climate change or to develop the technologies that might have done so? Might they bemoan that their foolhardy predecessors had deployed geoengineering and had recklessly made a bad situation even worse? Or might they be grateful that an earlier generation had deployed some geoengineering as part of a complex mix of policies that had been effective and had made a potentially bad situation bearable? These questions are unanswerable, not least because *future generations* are not a homogenous entity experiencing the same state of nature with a single worldview and set of cultural values. Yet, formulating near-term policy that responds to politicians' rhetorical concerns for the interests of future generations forces them to take a position on these questions, whether they do so explicitly or not.

From the perspective of future generations looking back, might they not wish that we had had the foresight to develop a portfolio of responses that would be as robust as possible given the widest range of plausible futures? The Golden Rule applies here. I propose that the standard by which we approach geoengineering be that we do for those in the future what we would wish they would have done for us had our temporal positions been reversed. On this basis, dismissing at the outset all forms of geoengineering at whatever scale would be justified only if we were supremely confident that our other responses were sufficient. As the figures in the next section suggest, there is little justification for such confidence. Arguments for dismissing geoengineering at the outset based on suppositions, in the absence of empirical evidence, that there are no circumstances in which it could reduce the overall risk from climate change, will be examined in detail in late chapters.

The numbers

If the 1,000PgC cumulative amount of GHG emissions is not to be breached by the end of the century, the global economy must undergo an extraordinary adjustment in the way it generates and uses power. Figure 1.2 illustrates one such pathway in which emissions peak in 2030 and gradually reduce to zero by 2100 at a rate that keeps us within the 1,000PgC target. There is an infinity of such pathways but self-evidently, the longer before the peak is reached or the higher the peak is, the steeper the subsequent reductions will need to be to meet the target. However, the problem is vastly more serious.

The trajectory of emissions is an analogue for economic activity. Figure 1.4 shows the close correlation between global GDP and carbon emissions over a period of almost two centuries with unconstrained access to fossil fuels. Two vital questions arise. First, what conditions must prevail for economic activity not to continue to increase more or less in line with the BAU projection, and secondly, to what extent might this activity be powered by non-CO_2 emitting fuels. There are currently billions of people in the world aspiring to raise their living standards to those in the developed economies, the trend towards urbanisation continues, the global population continues to grow with one UN estimate suggesting it might almost double by 2100 (*World Population to 2300* 2004, Table 1), and fundamentally, capitalism demands continued economic growth for its very survival. With these drivers it is difficult to see what will lead to significantly slower growth in economic activity and therefore, other things being equal, in energy demand. That being the case, the challenge for low and zero carbon energy sources is not simply to replace existing fossil fuels but also to supply all the future growth in energy needs. Figure 1.2 shows that within the next 85 years, assuming BAU growth in global GDP, the task for renewables is to supply the power represented by the dark grey area beneath the projected annual emissions curve, amounting to more than three times all the energy consumed to date since 1750.[15] Figure 1.5 shows the current extent of energy consumption fuelled by renewables. It also shows that never before has a new source of power

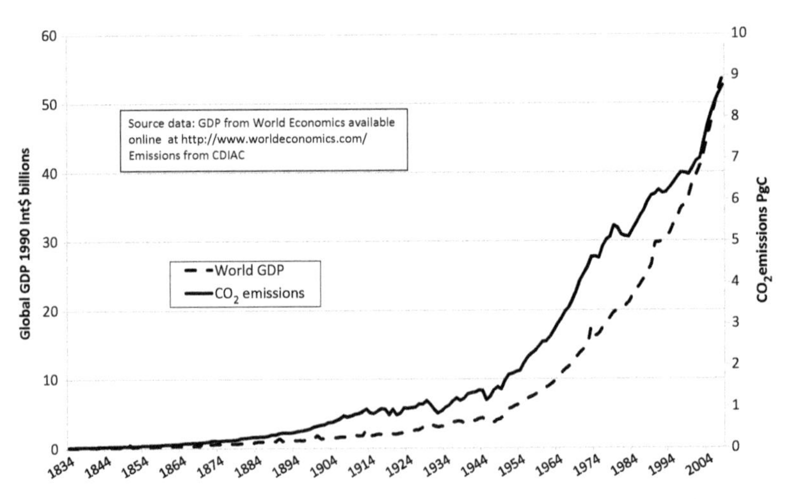

Figure 1.4 Correlation between Global GDP and CO_2 emissions 1834–2008. Source data: GDP from World Economics[16]; emissions from CDIAC.

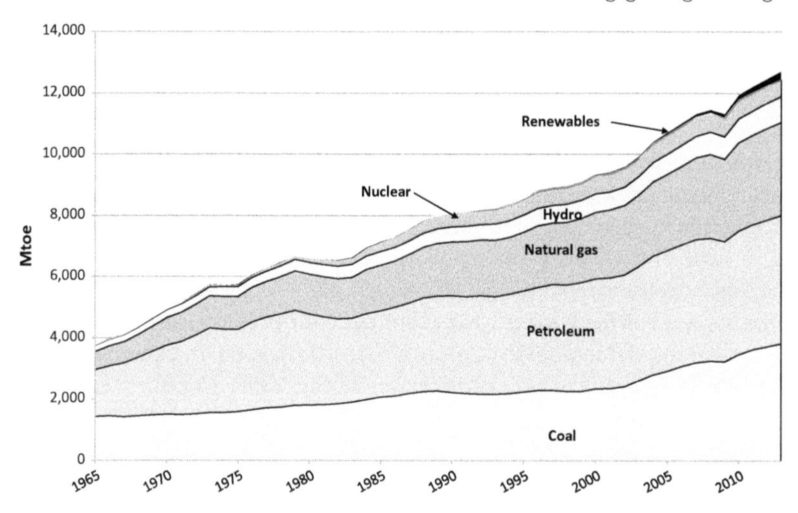

Figure 1.5 World primary energy consumption 1965–2013
Source: data from BP.[17]

replaced earlier sources; indeed, all energy sources are currently at, or close to, their historical maximums.

What is clear from these figures is that not breaching the 1,000PgC limit will be extremely challenging and therefore it would be prudent to consider what other routes at our disposal might lead to the mitigation of the deleterious effects of global warming that are expected to increase as we approach and pass that limit. Some geoengineering might be needed, indeed, as already noted, these figures assume at least ~1.5Pg of atmospheric carbon will be removed by CDR geoengineering every year from about 2070 onwards.

If there is any merit in this analysis, it follows that policymakers should begin to engage with geoengineering and this book examines how best this might be done. Policymakers are already familiar with the challenges of decision making under uncertainty. Those decisions are routinely made by municipal and national governments, corporate boards, charity trustees, university councils, and many other public and private collective entities. Even in the anarchic world of international relations there are bi- and multilateral treaties also representing decisions made under conditions of deep uncertainty. There is a great deal of accumulated human experience of a wide range of institutional structures and processes whose purpose is to make such decisions and to review and update, reform, revise and generally adapt them to changing current realities.

However, it seems to me that never before in human history has there been a need to make such decisions across such an extended temporal scale. Security, communications, health, transport, construction, agriculture, welfare – across the entire spectrum of political concerns, the actors are the current generation. The time horizon is short not only because the political cycles are rarely longer than five to ten years, but also, even more importantly, because there is no social policy where the outcomes do not begin to manifest over equally short periods. Climate change and geoengineering are not like that. Whatever policies are adopted to respond to climate change, the inertia in the global climate system is such that it could be decades before there is clear empirical evidence that its dangers are receding. Moreover, because of the absence of a controlled environment, in the form of a parallel planet unperturbed by policy interventions, it will always be challenging to identify the relative effectiveness of policies and even more so to identify which parts of a complex mix of policies have been more or less successful. Conventional policymaking is ill-equipped to formulate, implement and monitor policies with such an extended temporal horizon, making it difficult to attribute actual outcomes to policy initiatives.

The carbon cycle

Atmospheric CO_2 is part of a complex planetary scale carbon cycle. Some of the transitions within that cycle occur over short time scales; photosynthesis, for example, continuously converts atmospheric carbon into organic carbon throughout the day. Others occur over geologic time scales, for example the weathering of rocks that over millennia absorb CO_2 from the atmosphere to create carbonate minerals. The carbon cycle comprises sources, those processes that release carbon into the atmosphere, and sinks, those that remove it. If the sources exceed the sinks atmospheric concentrations of CO_2 increase. The burning of fossil fuels has created a vast new source of carbon that has unbalanced the equilibrium between sources and sinks that had been largely stable since the beginning of the Pleistocene more than one million years ago (Hansen and Sato 2011). The amount of the excess solar energy retained each year within the ecosphere is currently about twenty-five times the total of all the manufactured energy consumed by the entire global population.

The combination of the various natural processes that move carbon into and out of the atmosphere is such that the warming effect of atmospheric CO_2 endures for centuries (Archer *et al.* 2009). As a result, the anthropogenic atmospheric carbon emissions are continuously accumulating. This makes the task of emissions abatement as a route to averting dangerous climate change that much more challenging as each year of unabated emissions passes. As a consequence, adaptation to the changing environment and geoengineering are rising up the political agenda. The rapid increase in interest in geoengineering (Figure 1.3) is one of the responses to the continuing failure of the international community to implement timely and effective global GHG emissions abatement policies (Crutzen 2006).

The energy budget

Current geoengineering technologies are designed to manage the Earth's energy budget by influencing its incoming and outgoing energy flows (Keith

2000). There are only two ways to intervene in the global energy budget to reduce the energy retained within the planetary system, either we reduce net short wave radiation (SWR – principally light) arriving from outer space, or we increase long wave radiation (LWR – principally heat) transmission to outer space. These approaches are the basis for the now established division of geoengineering technologies into solar radiation management (SRM) and carbon dioxide removal (CDR) (Shepherd *et al.* 2009). The former addresses the SWR by a mix of reducing insolation and increasing the reflectivity of the Earth, technically known as its albedo; the latter seeks to reduce the amount of LWR retained by GHGs within the climate system, by reducing the concentration of atmospheric GHGs (by processes other than emissions abatement). A hybrid approach proposes reducing cirrus clouds to allow more LWR to escape.

CDR – *Reducing atmospheric carbon*

Emissions abatement apart, there is a limited number of ways in which atmospheric carbon concentrations can be reduced: either existing carbon sinks on land or in the oceans must be enhanced, or new carbon sinks created (Vaughan and Lenton 2011, 750). However, it is not sufficient for the carbon to be sequestered, it must also remain sequestered for an extended period. The permanence of a sink depends upon many factors, some of which are intrinsic, such as geological structures or biomass decay rates, and others exogenous such as site management. Lackner (2002, 200) argues that the lack of permanence and the demands of site management largely eliminate most biomass, soil, and many ocean disposal options as long-term sinks.

Keith *et al.*, on the other hand, are more sanguine about controlling leakage from geologic reservoirs and argue for direct air capture (DACS) of CO_2[18] as a means of continuously removing atmospheric CO_2 not only to restore any leakage but also to increase sequestration (Keith, Ha-Duong, and Stolaroff 2005). Leakage is a central concern for all sequestration methods and is affected by a range of factors specific to each storage site. It is a key intergenerational concern because, as Lackner suggests, sequestered carbon that returns to the atmosphere renders futile earlier progress in reducing GHG concentrations. Most importantly it would exacerbate the burden of climate change for future generations because they would have to deal not only with their own emissions and those accumulated in the atmosphere from previous generations, but also the premature release back into the atmosphere of the GHGs sequestered by earlier generations.

SRM – *Reducing net insolation*

It is solar energy that is the primary driver of the atmospheric temperature gradients that create the planet's climate systems. A reduction in insolation would not occur evenly over the planet even if the shielding effect of SRM is itself evenly distributed. This is so primarily because the sun's radiation falls at an increasingly acute angle towards the poles which reduces the cooling effect of the SRM. In addition, the same orographical features, cloud formations and aerosols that are responsible for local and regional variations in climate would also cause variability in the spatial effects of reduced insolation. It follows that the heterogeneous local and regional temperature changes from a global reduction in net insolation would result in complex changes to local and regional temperature gradients that in turn would drive potentially radically altered weather patterns. An SRM induced global cooling would not reverse, or even stop, the march of climate change, but would simply change the change, and not necessarily in universally benign ways. These dynamics will be examined in later chapters.

Prediction

Geoengineering is a special case of the more general public policy problem of decision making under uncertainty. Policymakers are generally aware of the importance of planning for the long term and powerful analytical tools now exist to assist in assessing risk and improving decision making in both the public and private spheres when confronted by deep uncertainty (Lempert, Popper, and Bankes 2003, xi). But rarely does the long term extend to the decades or even centuries that are relevant in decisions about climate change. Predicting policy outcomes becomes increasingly difficult as the temporal horizon extends into the distant future, and it is no less challenged by extending the spatial scope of policy from the local to the regional and ultimately to the global.

A Google search on 'false predictions' throws up a startling array of what, with the benefit of hindsight, seem extraordinary failures of foresight by informed people whom one might have expected to know better. Whether it is Einstein in the 1930s dismissing the possibility of civil nuclear power generation, Steve Ballmer, the CEO of Microsoft, confidently predicting in 2007 that the iPhone had 'no chance of gaining a significant market share', or Margaret Thatcher stating categorically a decade before becoming Prime Minster that there would be no female Prime Minister in the UK in her lifetime, the list is endless. Predicting distant future outcomes from near-term policy interventions in complex socio-technical systems are predictions of a different order. Such predictions are challenging not just because of their intractable uncertainty and complexity, but also because in many cases the outcomes will not even have been conceived of prior to their happening. When Watt and Boulton redesigned Newcomen's stream engine to make it commercially viable, no one then could have predicted the extraordinary impact the device would have thereafter. Very little of what we now take for granted as the material comforts of our modern world would have been possible without steam engines harnessing the power embedded in fossil fuels. But equally, little of it would have been conceivable by Watt or Boulton or their contemporaries. Yet-to-be-conceived futures defy prediction.

Prediction has been central to our growing understanding of the threats from climate change but if the extended deliberations of the UNFCCC teach us one thing, it is that the prospect, even the expectation, of distant threats is not a sufficiently powerful driver to galvanise urgent action at the necessary scale. While the UNFCCC pursues its elusive goal of global agreement on emissions abatement, governments at all scales, and businesses and private individuals are already changing their behaviours to reduce their emissions. Yet, Figure 1.1 shows that global emissions continue their inexorable growth. Whatever abatement is being achieved continues to be more than offset by increased emissions elsewhere. Clearly prediction has been insufficient, at least so far, to drive timely and effective emissions reduction policies. This opens the question as to what else is needed, or perhaps Gray is right and this is an example of our chronic incapability of learning from experience.

The role of prediction in climate policy is key. Had it not been for predictive methodologies we would not now be as aware as we are about the potential threats from climate change. Yet a central theme of this book is that it does not follow that simply because these methodologies have been effective in alerting us to the threats, that they are also sufficient, or possibly even necessary, for devising timely and effective policy responses.

Systems thinking

In parallel with the growing interest in geoengineering but almost entirely disconnected from it, from the late 1970s, Holling, Gunderson and Folke, amongst others (of whom more in Chapter 4) introduced into the world of ecology the pioneering work from the 1920s of an Austrian biologist, von Bertalanffy. He viewed organisms as complex systems in which the whole displayed emergent properties not apparent from any examination of its parts; a radical shift away from the reductionism introduced in the seventeenth century by Descartes and Bacon as they ushered in the Enlightenment. Intriguingly, this shift was occurring in the intellectually fertile inter-war years of the early twentieth century, as reductionism reached its zenith with the positivism of the Vienna Circle. von Bertalanffy's work has since been embodied in complex adaptive systems theory with applications across many disciplines from biology to corporate management.

This book brings together these two lines of thought; how can we intervene in the climate to avert dangerous climate change, and how might conceptualising the ecosphere as a complex adaptive system assist in that process. At its heart is the understanding and management of risk, uncertainty and surprise, and the policy implications of framing climate change as a situation to be continuously managed rather than as a problem to be solved. It embraces the complexity of the interdependent realms of the climate and society by examining how theories from the social sciences and systems theory, when applied to geoengineering, might help policymakers generate timely and effective near-term policies to avert long-term dangerous climate change. The evidence implied by the UNFCCC's continuing failure to secure any reduction in the growth of GHG emissions and the recent news that atmospheric concentrations have passed 400ppmv for the first time since *Homo sapiens* began roaming the planet, suggest that some policy alternatives are becoming increasingly urgent.

In one sense a proposition to adopt geoengineering does not raise questions different from any other policy issue faced by policymakers – what is the balance of benefits and burdens? And like other policy decisions, the benefits and burdens are not just those confined to the policy space itself, in this case responding to climate change, but also encompass a range of wider political factors such as its impact on power relations, public perception, and wealth distribution. But while the principles may be familiar, the practicalities are of a different order because of its great spatial and temporal extension and the dominant roles of risk, uncertainty and surprise. Even if the primary effects of geoengineering on, for example, global and regional temperatures or precipitation over long periods could be assessed with reasonable accuracy, their secondary and tertiary social and biodiversity effects will be path dependent and any predictions of these would be at best indicative and, quite probably, wildly wrong. If policymakers' response is to postpone action pending increased knowledge or reduced uncertainty, geoengineering offers immense opportunity for procrastination, an outcome entirely consistent with the procrastination to date by policymakers in relation to emissions abatement. As will be discussed in more detail in later chapters, more knowledge may actually increase uncertainty, and in a reductionist framing, more uncertainty promotes inaction.

Uncertainties about geoengineering come from many sources. First, the underlying climate science is still relatively young and there is much about the climate system that remains unknown. Satellites providing access to high quality empirical data at global scale for many climate variables are a relatively recent phenomenon, and there are many challenges associated with interpreting proxy data from the more distant past, for example tree rings dating back many hundreds of years and ice cores revealing a continuous record hundreds of thousands of years long. Second, the many different approaches to geoengineering introduce their own portfolio of risks not only related to their climatic effects, but also to the effectiveness and the safety of their engineering and delivery systems, and the social implications of the global scale industries necessary to support them.

Third, there is a wide range of spatial imponderables. There is only one global climate system and any interventions at scale that change it in one place will have effects elsewhere. Predicting those secondary effects, their nature, location, duration, and intensity, is subject to varying degrees of uncertainty. Fourth, and perhaps most challengingly, are the temporal considerations. Although the planet is large, it is finite. The future, on the other hand, is infinite which raises all manner of difficult questions. For example, how far into the future should we plan; will everyone in the future, near and distant, be of equal moral worth and is that equal to ours; while their vital needs such as nutrition, shelter, and security may not be much different from those of any human past or present, how should the largely unknowable and diverse ways in which they will be fulfilled as the decades and centuries pass be accounted for in our policy actions now? In addition, there is immense inertia in the climate system with effects taking many decades or even longer to manifest such that geoengineering interventions may have to be continued for decades or even centuries to deliver the desired results. These very long periods introduce considerable uncertainty in relation to the endurance of the institutions required to sustain them. The central concern of this book is to show how systems thinking might help to accommodate these uncertainties in active policymaking that reflexively adapts to increasing climate knowledge and changing social values.

Boundary issues

Geoengineering raises many interesting questions across many fields including ethics, national and international politics, economics, engineering and climate science. This book is primarily concerned with both the merit and the policy implications of geoengineering transitioning from an imaginary to a reality. This has obliged me to set aside many intriguing questions. I have not entered into the debate about whether the climate is changing or not and if so, the extent to which it is anthropogenic in origin, but have taken as given the mainstream view from the academy as set out in the IPCC reports. Except in order to distinguish geoengineering from other responses to climate change, I have not explored in any detail emissions abatement or adaptation.

A major issue in formulating geoengineering policy is the geopolitics of climate change. While I have made some reference to issues of power, the international relations dimensions of geoengineering have also been largely ignored except in relation to the process for governing geoengineering research (Chapters 8 and 9). Finally, I have ignored detailed questions about the economics of geoengineering, only addressing the manner in which economics handles risk, uncertainty and surprise to illustrate certain limitations of the discipline in coping with the spatial and temporal dimensions of climate change (Chapter 5).

Inside the boundary is an analysis of the dominant mindset at the root of the scientific method and how this manifests in climate science and geoengineering (Chapters 3 and 4). The development of complex adaptive systems theory by Folke, Holling, Gunderson and others, based on processes in ecology, are examined in some detail (Chapter 4). These analyses feed into an exploration of a range of approaches to problems with complex social characteristics. These include Rittel and Webber and their formulation of *wicked* problems that defy solution; Beck and his notion of *reflexive modernisation* as a way of understanding the social impacts of industrialisation and the pervasive spread of technology; Funtowicz and Ravetz's concept of *postnormal* science that argues that certain urgent high stakes problems require extensive public engagement if they are to be effectively confronted; and finally Cox's distinction between *problem-solving* and *critical theory* that, although developed in relation to international relations, has much to offer in understanding climate change and responses to it, challenges whose spatial extension places them firmly in the realms of the international.

I also refer to recent research concerning systemic resilience from within the social sciences by, for example, Evans and Stafford-Smith and others (Evans 2011; Stafford-Smith and Russell 2012). Additionally, the theoretical social science journal *Theory Culture & Society* published a special edition in 2010 dedicated to emerging ideas in social science relevant to climate change that illustrates an incipient shift away from conventional reductionism towards systems thinking.

The interests of proximate future generations, our children and grandchildren, are largely ignored in order to prevent short-term (in climate change terms) factors overwhelming benefits and burdens that may take a long time to emerge. There are also complex issues arising not only from relations between the current generation and future generations but also from relations between different future generations. There is a tendency to divide humanity into two distinct categories – the present generation and future generations – in which the diverse cultural traditions of present and past generations have become elided in the surreptitious homogenisation of future generations. In addition, there are important questions concerning the nature of any obligations we have to future generations and whether these are positive, placing upon us a duty to create a benign world for our descendants, or negative, in which our responsibilities are limited to not diminishing the opportunities they might enjoy. Despite all these intriguing questions, I have not discussed the ethical implications of geoengineering for intergenerational equity other than to note its emergence as a significant factor motivating responses to climate change and to reflect on how the indefinable heterogeneous cultural dynamics of indeterminate numbers of future generations, each comprising many local cultural traditions, might be addressed by a new geoengineering governance regime (Chapters 8 and 9).

Book structure

The book proceeds as follows:

Chapter 2 shows how the spatial and temporal dimensions of geoengineering emerge from the physical dynamics of the climate and climate change and identifies from within the climate system the physical roots of the main theme of this book, namely the socio-political challenges presented by geoengineering.

Chapter 3 explores the framing of climate change within the conventional Enlightenment paradigm and identifies some limitations in its capacity to address issues such as climate change that defy the processes of problematisation and solution that characterise the reductionist scientific method.

Chapter 4 introduces complex adaptive systems theory in relation to both the physical aspects of climate change and their social dimensions. It draws on parallels between wickedness, postnormal science, reflexive modernisation and critical theory, to argue that complex adaptive systems theory provides a unifying methodology for approaching the problematic of climate change and geoengineering, and overcomes many of the shortcomings of conventional reductionist methodologies.

Chapter 5 draws together ideas from the previous two chapters, and, focusing on the nature of risk, uncertainty and surprise in complex adaptive systems, examines the limitations of prediction as a methodology for formulating climate change and geoengineering policy. The outlines of a more effective approach to policy begin to emerge.

Chapter 6 illustrates by reference to the literature, the current reliance on prediction and the relative absence of systems thinking within public policy-making generally, and geoengineering more specifically. Nevertheless, it identifies from within the literature early signs of a reassessment of geoengineering in systems terms.

Chapter 7 considers the framing of geoengineering as a potential solution to climate change, whether as an alternative to emissions abatement or an emergency response. It concludes that geoengineering can only be an adjunct in a mixed portfolio of policies, and never a dominant policy.

Chapter 8 examines the nature of governance and regulation and applies this and insights from earlier chapters to formulate and critique the outline of a global geoengineering governance network.

Chapter 9 concludes by summarising the main arguments running through the book culminating in the proposal for the Geoengineering Governance Network. It further reflects on the capacity of emissions abatement technologies and changes in human behaviour to avert dangerous climate change, and concludes that the numbers indicate the increasing likelihood of some reliance on geoengineering. This provokes further reflection on our relations with distant future generations and arguments about whether or not it is morally acceptable to frame research into geoengineering as 'arming the future' against the possible ravages of dangerous climate change (Gardiner 2010). The book closes by discussing a plea, deeply buried in IPCC AR5, for a paradigm shift to promote systems thinking in place of conventional reductionist problem solving, as the core approach to the intractably complex social challenges presented by climate change and the responses to it, including geoengineering.

Conclusion

In a very real sense geoengineering turns the climate into an artefact. Galarraga and Szerszynski (2012) consider what this means and use the metaphors of the architect, the artisan and the artist to distinguish different kinds of 'making'. These stereotypes sit on a spectrum according to the extent to which the novelty of what is made emerges from skill in manipulating the materials or the creative spirit of the maker. They suggest that a climate made by geoengineering would fit more closely with the notion of the climate artist because 'making climates will change what the climate, the sky, and the weather are for us – their meaning, and place in human society'. They suggest that this brings with it the 'responsibility for and to the kind of world that geoengineering might bring into being' and conclude that we should ask 'what *kind* of god would we become if we started to make the climate' (emphasis in original). These are deeply interesting philosophical reflections but not ones, I suspect, that many policymakers will spend much time pondering.

The policy arguments for and against geoengineering will, I suggest, turn principally on an assessment of the extent to which it might reduce overall climate risks. Human activity does shape the planet's future but we are not, never have been and almost certainly never will be able to predetermine that future and devise policies that would deliver it according to our prior design. Uncertainty, surprise, error, and folly have an equal status in determining the future, and their effects are quintessentially unpredictable. If geoengineering is to become a significant part of the climate policy response mix, I shall argue that this can only happen as part of an heuristic process. This is not about being any kind of god, but rather about being human and working as best we can with circumstances and resources at our disposal. Geoengineering is not just a set of technologies but it is also part of a process whose objective is not to solve the problem of climate change, nor even to improve the likelihood of solving it, but rather a process whereby we and our successors can continually improve the likelihood of improving the likelihood that the threats from the rapidly changing climate recede and diminish. That is the essence of the structured culture of learning that, coupled with compassion, insight and wisdom, enables us to escape the clutches of Gray's paradox and benefit from experience. I shall argue that this requires a shift away from attempts to solve the problem of climate change to devising a portfolio of policies that is robust in the face of the widest set of plausible climate futures. Whether geoengineering has a place in that portfolio is too early to say, but it is also too early to reject.

Notes

1 John Gray discussing his new book *Soul of the Marionette: A Short Inquiry into Human Freedom* (2015) on BBC Radio 4 *Start the Week*, 6 April 2015.
2 For more information refer to the COP21 website at www.cop21.gouv.fr/en.
3 In these opening paragraphs I use the words weather and climate somewhat loosely. In meteorology, climate is defined as the weather averaged over a period. The UK Met Office defines 'climate' as follows: 'Climate, which comes from the Greek *klima* meaning "area", usually refers to a region's long-term weather

patterns. This is measured in terms of average precipitation (i.e. the amount of annual rainfall, snow etc), maximum and minimum temperatures throughout the seasons, sunshine hours, humidity, the frequency of extreme weather, and so on. These days, "climate change" usually refers to global climate change, or long-term variations in the planet's average temperature. But it can also be used more generally to mean local and regional changes in weather patterns. The global average temperature is influenced by many interacting systems which, together, we call the climate system. But, of course, global climate change also affects regional climates.' www.metoffice.gov.uk/climate-guide/climate/, accessed 23 September 2015.

Climate is an intellectual construct; it is the weather that we actually experience.

4 Common alternatives to *geoengineering* are *climate engineering* and *climate remediation*. Those favouring *climate engineering* prefer to emphasise the specifically climatic rather than more general *geo*-aspects of planet Earth as a whole. *Climate remediation* is intended to connote some form of restoration of the climate to a prior more benign state. The subtle nuances intended by their respective champions are of little consequence in this book and I shall treat these terms as synonymous.

5 Available online at http://unfccc.int/resource/docs/convkp/conveng.pdf/, accessed 13 March 2011.

6 The international community has adopted the term *mitigation*, as used in the UNFCCC, to mean emissions abatement. Unfortunately when this document was drafted in the early 1990s, neither adaptation nor geoengineering were on the agenda, and the only way of mitigating the effects of climate change then considered was emissions abatement. However, as time has passed and both adaptation and geoengineering have been included as additional approaches to addressing the effects of climate change, some linguistic confusion has arisen because both of these are also ways of mitigating the effects of climate change. In this book I shall avoid this confusion by using the term *emissions abatement* to refer to what the UNFCCC calls mitigation, and using the word *mitigation* in its more general usage to refer to any behaviour that mitigates, or reduces the negative effects of, some undesirable condition.

7 Innovation and economic development are and have always been unevenly distributed throughout humanity (Malm and Hornborg 2014) and not all humanity can be implicated equally in the encroachment of civilisation upon the environment. Nor is *Homo sapiens* the only species to have had climate changing impacts on the environment, for example the change from an anoxic to an oxidising atmosphere was largely brought about by cyanobacteria some 2.5 bn years ago (Sessions *et al.* 2009). Nevertheless, the current problem of anthropogenic emissions induced climate change is, by definition, anthropogenic. The 'theorization of culture and power' that Malm and Horburg hold to be the concern of social science can unpick the fine details of the diverse contributions of different segments of humanity in this process without undermining the fact that anthropogenic climate change is induced by humanity, even if not by them all.

8 www.elysee.fr/documents/index.php?lang=fr&mode=cview&cat_id=2&press_id=2754/, accessed 21 July 2009 (transl: Future generations will not forgive us [if we fail to reach a worldwide agreement on global warming]).

9 www.klima-luegendetektor.de/2008/12/10/angela-merkel-cdu-das-klima-chamaleon/, accessed 6 September 2009 (transl: The fight against global warming is a question of the survival of mankind).

10 Refer to www.aip.org/history/climate/internat.htm for a comprehensive history of the development of the international awareness and response to climate change, from the formation of the International Meteorological Organization in 1879 through the post-World War II resurgence in interest with the International Geophysical Year in 1957/58, the Villach conference in 1985 that provided the

first warnings of global warming from accumulating atmospheric greenhouse gases, closely followed by the 1987 Montreal Protocol banning ozone depleting substances, the 1988 establishment of the IPCC and the Toronto Conference highlighting the role of greenhouse gases, and the 1992 Rio Earth Summit and beyond.

11 Fossil fuel and cement production emissions data from CDIAC at http://cdiac.ornl.gov/.

12 Data from CDIAC Trends, available online at http://cdiac.ornl.gov/trends/landuse/houghton/houghton.html. Data covers the period from 1850 to 2005 from which Houghton concludes that aggregate CO_2 emissions from land use changes during that period were 156PgC. He also infers that global fluxes amounted to ~1.5$PgCyr^{-1}$ thereafter. This yields aggregate emissions from land use change for the period from 1850 to 2012 of ~170PgC.

13 Normativities surrounding what might be considered *sufficient knowledge* and *acceptably low risk* are considered in some detail later, particularly in Chapter 5.

14 This 'watching brief' status was confirmed in a personal conversation (19 September 2012) with Jolene Cook at the UK Department of Energy and Climate Change who, at the time, was the civil servant responsible for advising Ministers on geoengineering.

15 It is assumed that in the fossil fuel era, energy demand has been more or less proportional to CO_2 emissions.

16 Data available online at http://www.worldeconomics.com/, accessed 8 April 2015.

17 www.bp.com/statisticalreview, accessed 23 March 2015.

18 There is some difference of opinion about whether carbon capture and storage (CCS) when used in conjunction with power generation should be regarded as emissions abatement or geoengineering. Some argue that processes that prevent CO_2 emissions are abatement whereas those that involve the capture of CO_2 already emitted into the atmosphere (DACS) are geoengineering. This difference relates only to the capturing of the CO_2 – the process of sequestering the captured CO_2 is identical in both cases. I shall follow Vaughan and Lenton (2011) in treating CCS as abatement and DACS as geoengineering.

References

Archer, D., M. Eby, V. Brovkin, A. Ridgwell, L. Cao, U. Mikolajewicz, K. Caldeira, K. Matsumoto, G. Munhoven, and A. Montenegro. 2009. 'Atmospheric Lifetime of Fossil Fuel Carbon Dioxide'. *Annual Review of Earth and Planetary Sciences* 37(1): 117.

Asilomar Scientific Organizing Committee. 2010. 'The Asilomar Conference Recommendations on Principles for Research into Climate Engineering Techniques'.

Bala, G. 2009. 'Problems with Geoengineering Schemes to Combat Climate Change'. *Current Science* 96(1): 41–48.

Barrett, S. 2014. 'Solar Geoengineering's Brave New World: Thoughts on the Governance of an Unprecedented Technology'. *Review of Environmental Economics and Policy* 8(2): 249–269. doi:10.1093/reep/reu011.

Bracmort, Kelsi, Richard K. Lattanzio, and Emily C. Barbour. 2010. 'Geoengineering: Governance and Technology Policy'. In Congressional Research Service, Library of Congress.

Brovkin, Victor, Vladimir Petoukhov, Martin Claussen, Eva Bauer, David Archer, and Carlo Jaeger. 2009. 'Geoengineering Climate by Stratospheric Sulfur Injections: Earth System Vulnerability to Technological Failure'. *Climatic Change* 92(3): 243–259. doi:10.1007/s10584–10008–9490–9491.

Bunzl, Martin. 2009. 'Researching Geoengineering: Should Not or Could Not?' *Environmental Research Letters* 4: 045104.
Cicerone, R. J. 2006. 'Geoengineering: Encouraging Research and Overseeing Implementation'. *Climatic Change* 77(3): 221–226.
Crutzen, Paul J. 2006. 'Albedo Enhancement by Stratospheric Sulfur Injections: A Contribution to Resolve a Policy Dilemma?'. *Climatic Change* 77(3–4): 211–220. doi:10.1007/s10584-10006-9101-y.
Durand, Rodolphe, and Eero Vaara. 2006. *A True Competitive Advantage? Reflections on Different Epistemological Approaches to Strategy Research*. Chambre de Commerce et d'Industrie de Paris. www.hec.edu/heccontent/download/4765/130926/version/2/file/CR838.pdf.
Evans, J. P. 2011. 'Resilience, Ecology and Adaptation in the Experimental City'. *Transactions of the Institute of British Geographers* 36(2): 223–237.
Fleming, James Rodger. 2010. *Fixing the Sky: The Checkered History of Weather and Climate Control*. Columbia University Press.
Galarraga, M., and B. Szerszynski. 2012. 'Making Climates: Solar Radiation Management and the Ethics of Fabrication'. In *Engineering the Climate*, edited by Christopher J. Preston, 221–235. Lexington Books.
Gardiner, Stephen. 2010. 'Is "Arming the Future" with Geoengineering Really the Lesser Evil?' In *Climate Ethics: Essential Readings*, edited by Stephen Gardiner, Simon Caney, Dale Jamieson, and Henry Shue, 285–312. OUP USA.
Gardiner, Stephen. 2011. 'Some Early Ethics of Geoengineering the Climate: A Commentary on the Values of the Royal Society Report'. *Environmental Values* 20(2): 163–188. doi:10.3197/096327111X12997574391689.
Gray, John. 2015. *The Soul of the Marionette: A Short Enquiry into Human Freedom*. Allen Lane.
Hamilton, Clive. 2013. 'Ethical Anxieties About Geoengineering'. In *Ethics and Emerging Technologies*, edited by Ronald L. Sandler, 439. Palgrave Macmillan.
Hansen, J. E., and M. Sato. 2011. 'Paleoclimate Implications for Human-Made Climate Change'. Arxiv Preprint arXiv:1105.0968.
IPCC. 2014. *Climate Change 2013 – The Physical Science Basis: Working Group I Contribution to the Fifth Assessment Report of the Intergovernmental Panel on Climate Change*. Cambridge University Press.
Keith, David W. 2000. 'Geoengineering the Climate: History and Prospect'. *Annual Review of Energy & the Environment* 25(1): 245.
Keith, David W., Minh Ha-Duong, and Joshuah K. Stolaroff. 2005. 'Climate Strategy with CO2 Capture from the Air'. *Climatic Change* 74(1–3): 17–45. doi:10.1007/s10584-10005-9026-x.
Lackner, K. S. 2002. 'Carbonate Chemistry for Sequestering Fossil Carbon'. *Annual Review of Energy and the Environment* 27(1): 193–232.
Lempert, R. J., Steven W. Popper, and Steven C. Bankes. 2003. 'Shaping the Next One Hundred Years'. RAND Corporation. www.rand.org/pubs/monograph_reports/MR1626.html.
Malm, Andreas, and Alf Hornborg. 2014. 'The Geology of Mankind? A Critique of the Anthropocene Narrative'. *The Anthropocene Review*, January, doi:10.1177/2053019613516291.
Marshy, Leila, ed. 2010. 'Geopiracy: The Case Against Geoengineering'. ETC Group. www.etcgroup.org/upload/publication/pdf_file/ETC_geopiracy2010_0.pdf.

Moore, Henrietta. 2015. 'The End of Development'. BBC Radio 4. www.bbc.co.uk/programmes/b054pqv8.
NAS. 2015. 'Climate Intervention: Reflecting Sunlight to Cool Earth'. National Academy of Sciences. www.nap.edu/catalog/18988/climate-intervention-reflecting-sunlight-to-cool-earth.
Oldham, P., B. Szerszynski, J. Stilgoe, C. Brown, B. Eacott, and A. Yuille. 2014. 'Mapping the Landscape of Climate Engineering'. *Philosophical Transactions of the Royal Society A: Mathematical, Physical and Engineering Sciences* 372(2031): 20140065. doi:10.1098/rsta.2014.0065.
Pierrehumbert, Raymond T. 2015. 'Climate Hacking Is Barking Mad'. *Slate*, February 10. www.slate.com/articles/health_and_science/science/2015/02/nrc_geoengineering_report_climate_hacking_is_dangerous_and_barking_mad.html.
Preston, Christopher J., ed. 2012. *Engineering the Climate: The Ethics of Solar Radiation Management*. First Edition. Lexington Books.
Robock, Alan. 2008. '20 Reasons Why Geoengineering May Be a Bad Idea'. *Bulletin of the Atomic Scientists* 64(2): 14–59.
Sessions, Alex L., David M. Doughty, Paula V. Welander, Roger E. Summons, and Dianne K. Newman. 2009. 'The Continuing Puzzle of the Great Oxidation Event'. *Current Biology* 19(14): R567–574. doi:10.1016/j.cub.2009.05.054.
Shepherd, John, Ken Caldeira, P. Cox, J. Haigh, David W. Keith, B. Launder, Georgina Mace, G. MacKerron, J. Pyle, and Steve Rayner. 2009. 'Geoengineering the Climate: Science, Governance and Uncertainty'. The Royal Society.
Stafford-Smith, M., and L. Russell. 2012. 'A Resilience View on Reframing Geoengineering Research and Implementation'. *Carbon* 3(1): 23–25.
Stilgoe, Jack. 2015. *Experiment Earth: Responsible Innovation in Geoengineering*. Routledge.
Szerszynski, Bronislaw, Matthew Kearnes, Phil Macnaghten, Richard Owen, and Jack Stilgoe. 2013. 'Why Solar Radiation Management Geoengineering and Democracy Won't Mix'. *Environment and Planning* A 45(12): 2809–2816. doi:10.1068/a45649.
Thatcher, Margaret. 2002. *Statecraft*. HarperCollins.
Tuchman, Barbara Wertheim. 1984. *The March of Folly: From Troy to Vietnam*. First Edition. Alfred A. Knopf.
United Nations. 2004. *World Population to 2300*. United Nations Publications.
Vaughan, N. E., and T. M. Lenton. 2011. 'A Review of Climate Geoengineering Proposals'. *Climatic Change* 109(3): 745–790. doi:10.1007/s10584–10011–0027–0027.
Virgoe, John. 2009. 'International Governance of a Possible Geoengineering Intervention to Combat Climate Change'. *Climatic Change* 95(1): 103–119. doi:10.1007/s10584–10008–9523–9529.

2 Geoengineering – the technologies and their 'times'

Time and space provide the dimensions that define and constrain the unfolding of climate change and any attempt to mitigate its effects. If the science and the policy that drive these attempts are not sensitive to the temporal and spatial dimensions of climate change, they will succeed only by pure chance, and are more likely to aggravate than improve the situation. In this chapter these aspects of geoengineering are examined in order to provide the foundations for the discussions in later chapters about the intersection of uncertainty and complexity in climate change policy.

Geoengineering – a summary of the current technology options

There are a dozen or more technologies currently under consideration to intervene intentionally in the global climate to avert dangerous climate change. I shall briefly describe the most discussed technologies. In the first instance I use the established SRM/CDR classification but then I point to various aspects of their risk/benefit profiles that undermine the common view that while CDR is benign, SRM is risky, and moreover, that in general terms, CDR is less ethically objectionable than SRM.[1] This section is based largely on material from the Royal Society (Shepherd *et al.* 2009), the Copenhagen Consensus on Climate (Bickel and Lane 2009) and Vaughan and Lenton's update (2011) of an earlier review of geoengineering technologies. These sources are surveys of the technology concepts currently in circulation; they provide extensive references to original sources that are not repeated here.

It is first necessary to appreciate that neither SRM nor CDR is a quick-fix for climate change and that both introduce major intergenerational considerations whose nature depends critically on the role ascribed to geoengineering in the policy mix. Several SRM technologies could reduce GMST by an appreciable amount within the span of two to five years if deployed at sufficient scale (Crutzen 2006; Govindasamy and Caldeira 2000). However, the benefits of SRM only endure while the reflectors are in place and therefore if SRM were to be a significant component of our response to climate change, it would have to be deployed continuously for many decades, and possibly even centuries. CDR technologies, on the other hand, would have no material impact on GMST in the short term but could provide an important cumulative benefit if continued over many decades and centuries (Caldeira 2009, 4.1.3.2; Lempert and Prosnitz 2011, 43–44).

Quite apart from the future development of nascent and as yet unconceived SRM and CDR technologies, there can be little doubt that the operational details of those already identified will undergo radical development in the light of further theoretical and empirical research. In surveying these technologies, the reader should be aware that the incorporation of any of them into the mix of responses to climate change would imply a multigenerational commitment during which time they would almost certainly evolve in ways difficult to predict. At this early stage it is important, therefore, not to dismiss technologies too readily merely because of current technical challenges that might well be addressed in the future.

In many cases the geoengineering technologies are the idea of a single individual or small academic team, and the almost complete absence of any empirical data places them firmly in the realm of the conceptual. Even in those few cases where there is some geoengineering observational data (e.g. marine cloud brightening and ocean iron fertilisation), the data is thin and inconclusive (Blackstock *et al.* 2009; Victor *et al.* 2009).[2] Nevertheless, in every case the basic chemistry and physics underpinning the technology is relatively uncontested and for some is supported by evidence from volcanic eruptions. However, implementation at a climatically significant scale is usually an unresolved problem and in no case is it clear how the programmes would be financed at full deployment scale.

Only the briefest explanation of each technology will be given here in order to give the reader a basic understanding of the range of technologies involved and their spatial and temporal characteristics. I describe their principal mode of action and the major advantages and disadvantages of each; economic considerations are mostly ignored.

CDR technologies

These methods all operate by directly removing CO_2 from the atmosphere to reduce atmospheric CO_2 concentrations thereby reducing their capacity to capture LWR emanating from the Earth's surface. This reduces the radiation of LWR back to the Earth's surface, and allows it to escape into outer space. All these technologies suffer from two significant disadvantages. First, to be significant, their climatic impact would require deployment at scale over many decades. If CO_2 emissions continue to increase at their recent rate, CDR may be incapable of averting the dangerous climate change referred to by the UNFCCC (Caldeira 2009, 4.1.3.2; Lempert and Prosnitz 2011, 43–44). Second, the scale at which they would have to be undertaken would have major ecological and economic consequences (Lackner 2002, 198). The ecological impacts would arise from their widespread intrusion in a range of human activities from, for example, land use change, mining, transport and distribution, and the production of waste. The economic impacts would arise from extensive competing demands on resources. While CDR might produce a great deal of economic activity that would be beneficial for some, it would also consume vast amounts of natural resources, and with little marketable output those bearing the costs might find them burdensome given their other more pressing social priorities. The problem of scale is illustrated by the example of carbon sequestration by biochar (see Appendix). This computation, despite considerable uncertainty in valuing some of the variables, suggests that to sequester $1PgCyr^{-1}$ in the form of biochar would require a land area half the size of the USA on which to grow the timber feedstock. This timber would weigh almost half of current global annual coal production and would require shipping containers equivalent to more than half of current global annual port container traffic. It would also require 13 biochar plants to be built every week for 50 years to reach a capacity of just $1PgCyr^{-1}$ of the $15PgC_{eq}yr^{-1}$ of current GHG emissions.

Many of the challenges presented by CDR at climate-changing scale might be more easily overcome if the captured CO_2 could be converted into a feedstock for some profitable good or service, particularly if it were to provide long-term sequestration rather than short-term churning. No such opportunities are presently available; current industrial uses of CO_2 represent a minute fraction of annual emissions and the vast bulk of these are released back into the atmosphere within months (Mazzotti *et al.* 2005, Table 7.2).

Land use, land cover changes (LULCC)

Land use emissions can be reduced by less tilling of the soil and less land clearance for agriculture. Land use changes can also drawdown more atmospheric CO_2 through photosynthesis by afforestation (creating new forests), reforestation (replacing felled forests) and avoiding deforestation. Advantages: technically straightforward and scalable. Disadvantages: limited storage capacity, low permanence of carbon sink, potential conflicts with other land uses in particular for food production; secondary impacts on water availability, fertiliser run-off and biodiversity (Vaughan and Lenton 2011, 750).

Biomass with carbon capture and storage (BECCS)

This process entails capturing the CO_2 emissions from biofuel combustion and sequestering them in geological formations. It may also entail growing dedicated biofuel crops. Advantages: biomass production technically straightforward and scalable. Disadvantages: same land use issues as for LULCC, scalable technologies for end-of-pipe CO_2 sequestration are proving difficult to develop; and leakage risk from transport and sequestration of captured CO_2.

Biomass and biochar

The principle here involves burying biomass underground where it will slowly decay. Alternatively, by first pyrolising the biomass (combustion in low oxygen atmosphere to produce charcoal), the biochar residue becomes a long-lasting carbon sink. Additionally, the pyrolisation can be used for electricity generation (emitting about 50 per cent of its carbon). Advantages: technically straightforward and scalable, improvements to soil quality from some added biochar. Disadvantages: same land use issues as for LULCC, unknown effects of radically increasing the amount of charcoal in soils (see Appendix).

Enhanced weathering on land and at sea

Silicates are the basis of most rocks and react with atmospheric CO_2 to form carbonates that are more or less permanently stable carbon sinks. Over geologic time this process is extremely important in the carbon cycle. Technologies have been proposed for accelerating this process, involving carbonate deposition on land and at sea. Advantages: permanence and safety of stored carbon, virtually unlimited storage capacity. Disadvantages: technologically demanding at scale and requiring establishment of global mining industry probably greater in size than existing fossil fuel extraction; ocean sequestration effectiveness reduced by ocean acidification.

Chemical air capture and carbon sequestration

This process, sometimes referred to as artificial trees or direct air capture and sequestration (DACS), centres on technologies that use chemical reactions to scrub CO_2 from the atmosphere and then sequester it in gaseous or liquid form in geological formations. Advantages: air capture is scalable with few known side effects; large availability of geologically suitable sequestration sites. Disadvantages: extremely long term (centuries to millennia) requirement for monitoring and maintenance of storage sites; high energy requirement for carbon capture, volatility of sequestered carbon and risk of leakage; storage in deep ocean sites is technically demanding and is subject to unknown variables in deep ocean dynamics (Lackner 2002; Vaughan and Lenton 2011, 760); extremely demanding of local water resources.

Ocean sink

The oceans are an important carbon sink, and although almost in balance with a net uptake of only $2PgCyr^{-1}$, the annual fluxes are substantial with some $160PgCyr^{-1}$ being exchanged between the atmosphere and the oceans (IPCC 2014, Fig. 6.1). Ideas for enhancing the ocean as a carbon sink involve a range of approaches to transport carbon to the deep ocean. This carbon may be in the form of hard carbonates (e.g. shells and marine skeletons) or in soft organic tissue, or in dissolved CO_2. Various methods have been proposed from increasing the growth of phytoplankton by fertilising the ocean with scarce nutrients, to intervening in the transport systems that move large masses of ocean water between the surface and the deep ocean. Advantages: almost limitless storage capacity. Disadvantages: some methods are ineffective, others are technically demanding; all involve changes to ocean chemistry and ecosystems with potentially undesirable side effects (Vaughan and Lenton 2011, 753 et seq.).

SRM technologies

These technologies operate by preventing SWR from entering the atmosphere or reflecting it directly back into outer space before it has been converted into retained heat energy. If deployed at sufficient scale, they have the potential to reduce global temperatures within a matter of a few years, and possibly even in months. However, there appears to be an inverse relationship between effectiveness and risk (Rayner 2010). A central weakness of all SRM technologies is that they offer no means of reducing ocean acidification. Indeed, if SRM caused complacency in emissions abatement, ocean acidification would be exacerbated.

Human settlement and desert albedo

This requires that roof and other surfaces in the built environment be painted white or covered with reflective materials. A variation on this theme is to install vast reflectors in the desert. Advantage: technically simple, ample desert capacity. Disadvantages: limited capacity within the built environment that may have local urban heat island benefits but not significant global effects, reflective surfaces would require high maintenance, especially in desert settings.

Grassland and crop albedo

This extends the previous idea to rural environments by sowing more reflective crops and thereby reflecting more sunlight back into space. Advantage: conceptually attractive since crops are grown across vast areas of the planet. Disadvantages: unknown environmental and biodiversity effects from changed photosynthetic dynamics given the scale at which modified crops would need to replace existing crops, biodiversity effects from extensive monoculture.

Marine cloud brightening

By reducing the size of the particulates that seed moisture droplets in clouds, the clouds can be made denser and more reflective. To reverse a doubling of atmospheric CO_2, the technology envisages a fleet of 1,500 unmanned ships spraying fine jets of sea-water. Advantages: scalable. Disadvantages: unknown weather effects from significantly reduced local net insolation (due to higher cloud albedo) and concomitant effects on photosynthesis that could reduce marine biotic take up of CO_2, and on precipitation from reduced evaporation.

Stratospheric aerosols

The proposal is to replicate the cooling effect of volcanic sulphate aerosol emissions into the stratosphere. A wide variety of delivery mechanisms has been considered but the most studied is to inject aerosols directly into the lower stratosphere (~15km above the Earth's surface, slightly above the cruising height of most commercial passenger jet aircraft) where they would diffuse sunlight entering the atmosphere, reflecting a portion of it back into outer space. The undesirable effect that this might have on stratospheric ozone has led some to consider injecting the aerosols into the troposphere (Tilmes *et al.* 2009). In either case, the aerosol injections would need to be continuous because the aerosols have a relatively short residence. Advantages: technically feasible and scalable, increased uptake of CO_2 through photosynthesis in response to increased diffuse light, rapid effect. Disadvantages: depletion of stratospheric ozone, possible disruption of major weather systems on which many millions depend for their food supplies, e.g. Asian and African monsoons – although there is evidence that climate change is already having this effect (Auffhammer, Ramanathan, and Vincent 2012), reduction in efficiency of solar energy technologies, unknown impacts from large scale increased efficiency of photosynthesis from diffused light, no impact on ocean acidification.

Space-based reflectors

Described by Schneider as Buck Rogers schemes (Schneider 2008), reflective surfaces would be launched into space either in near-Earth orbits or at the Lagrange L1 point approximately 1 million miles from Earth, where the gravitational pull of the sun and Earth is cancelled out. These mirrors would prevent a portion of the sun's radiation from reaching the Earth's atmosphere. Advantages: scalable. Disadvantages: technically demanding and a long time before any feasible deployment; for Lagrange L1 deployment inability to control or switch off in the event that the undesirable side effects become untenable, maintenance required to control cosmic drift (Shepherd *et al.* 2009, 37).

Cirrus cloud stripping

High level cirrus clouds absorb and re-emit LWR but allow SWR to pass through. By reducing the amount of cirrus clouds more of LWR would be allowed to escape to outer space. This is a hybrid technology that is neither SRM nor CDR. The intervention in cloud formation makes this proposal more akin to SRM but its effect is similar to CDR (Mitchell and Finnegan 2009).

Spatiality and temporality of geoengineering technologies

In the context of climatology, spatiality refers to the physical distribution of climate variables across all scales from the local to the global. This physical spatiality encompasses both the meteorological phenomena, for example, temperature, precipitation, and atmospheric pressure, and the physics and chemistry that drives changes in them over short and long time scales. But it also has a social dimension. Much human behaviour is driven by the spatial dynamics of the climate, for example the different clothing and dietary habits of the Inuit in the Arctic and the Hutu in equatorial Rwanda. A central issue of the politics of climate change concerns the many different ways that communities will be affected according to their geographic location. Climate change and its effects do not recognise state borders, nor do they discriminate between rich and poor (although the rich may be better able to cope with them), nor between those communities whose early industrialisation makes them historically more responsible for climate change than those in less developed countries (e.g. Adger 2003; Hulme 2010; Jasanoff 2010).

Whereas the concept of globality in the social sciences generally refers to practices that arise locally but are replicated globally, geoengineering is possibly unique in human history in that it refers to an activity that, wherever and however it is implemented, is conceived as a global-scale intervention, albeit that its effects may vary spatially. Global impact is an essential feature of geoengineering and requires either high impact technologies to be deployed in a relatively limited number of locations (e.g. vast arrays of mirrors in a few desert locations or a few sites from which aerosols are lofted into the stratosphere), or low impact technologies more widely distributed around the globe (e.g. extensive use of highly reflective crops, afforestation, DAC, BECCS, biochar or marine cloud brightening).

The Royal Society (Shepherd *et al.* 2009) proposed eight criteria by which to evaluate geoengineering projects: legality, effectiveness, timeliness, impacts, costs, funding support, public acceptability and reversibility. Each of these criteria displays both temporally and spatially specific attributes. What is legal in one jurisdiction at a given point in time, may not be in another or at a later date. What is effective in mitigating the effects of climate change in one location may produce undesirable outcomes in another, and these may vary through time. Timeliness is entirely culturally dependent; the urgency of addressing climate change already palpable for inhabitants of the small island Oceanic states is radically different from the more measured pace being taken in most of the developed world.[3]

The capacity to understand, verify, and stop or modify already deployed geoengineering will be different in different locations and will certainly change through time with the accumulation of further knowledge. Costs are also likely to be distributed unevenly, and both their amount and distribution may be radically different in the future. Funding support will also vary between nations and this may change in the future as the tectonic plates of the global economy shift. What is publicly acceptable in almost any domain of public policy is already demonstrably spatially diverse, both within and between countries, and there is no reason for attitudes and responses to climate change to be any different. Finally, what is considered reversible is also likely to be spatially dependent according to the precise manner in which the effects of geoengineering arise in different locations. It also seems reasonable to suppose that as understanding and engineering capacity develop in the future, effects that might once have been thought irreversible become reversible, at least to some extent.

The Royal Society observed that none of the proposed methods met all their criteria and concluded it would be 'necessary to balance [their] different properties against one another, and this is bound to raise differences of opinion' (Shepherd *et al.* 2009, 19). These differences of opinion are rooted in unequal spatialities and temporalities. Much of the problematic of geoengineering would disappear if the problems it sought to resolve, the ways in which it sought to resolve them and the impacts of these efforts affected everyone everywhere equally, and that those bearing the burdens of dealing with climate change were those who had caused it and who would also reap the benefits of their efforts to combat it. If that were the case the challenge would merely be to develop the most cost effective technologies; intra- and intergenerational conflicts of interest would not surface. But the spatial heterogeneity of values and interests and their spatially distinct development through time demand continuing global engagement in the assessment of the balance of burden and benefit from geoengineering. In later chapters I shall examine the challenges in making such assessments but in the balance of this chapter, I examine their temporal dimension in more detail.

The 'times' of geoengineering

Geoengineering has a range of times associated with different aspects of its various technologies. These include *readiness time* – the time needed to bring them to a state of deployment readiness; *impact time* – the time it takes to achieve their desired (and unintended and perhaps undesirable) climate effects; *deployment time* – the period during which their deployment would need to be continued; *termination time* – the time it would take to switch off a deployed geoengineering activity; *cessation time* – the time elapsed after termination during which the cooling effect of the geoengineering continues; and *permanence* refers to the period of time before sequestered carbon returns to the atmosphere either through natural processes or the failure of either the social or engineering systems necessary to contain it. These temporalities will be different for different geoengineering technologies. For example, it will take considerably more time to have a space-based shield of mirrors ready for deployment than to inject sulphates into the stratosphere (Schneider

2008); and sulphating the stratosphere will reduce GMST much more rapidly than any combination of CDR technologies. Each of the times has somewhat different intergenerational consequences.

Readiness time

All the CDR methods depend upon relatively simple chemistry but entail immense technical and financing problems in being scaled to global effectiveness (Shepherd *et al.* 2009). It is difficult to be precise about these timescales because until empirical research begins in earnest, there remain too many uncertainties. Most geoengineering technologies could probably be made ready for deployment at scale within 50 years, and possibly sooner if supported by the political will (Bickel and Lane 2009, 16–17). More distant generations would be better placed to decide in favour of geoengineering (or otherwise), because they will have a longer experience of climate change on which to draw, and will also by then have a better understanding of the physics and chemistry of the climate and its interdependency with humanity. However, the longer GHG emissions continue to grow, the more urgent the climate problem will become for those in the future. That urgency may limit their options to the faster acting, more risky technologies. Nobel economics laureate Kenneth Arrow makes the point in his paper on uncertainty and reversibility in environmental matters (although written long before climate change was an issue) that 'the expected benefits of an irreversible decision should be adjusted to reflect the loss of options it entails' (1963, 319). In the context of climate change and geoengineering, this remark might be recast, with equal effect – the expected cost of irreversible *indecision* should be adjusted to reflect the loss of options it entails.

Deployment time

The duration of any geoengineering deployment will be a continuous concern for future generations. They will be confronted by the multiple tasks of continuing the deployments, assessing their impacts, fine tuning the specifics of the deployment and establishing and re-establishing climate stabilisation targets. It cannot be assumed that the desired goals of geoengineering will remain fixed through time. If ways are found to make region-specific geoengineered climates, rather than a one-size fits all single global warming target, future generations will be continuously faced with the need to assess and reassess those regional targets and the conflicts of interest they entrain. However, this applies even without region-specific geoengineering because of the heterogeneous spatial effects of all forms of geoengineering. This process implies some form of global governance of a form that currently does not exist. The governance regime itself will be in constant evolution to reflect the socio-political demands of the day as well as the changing environmental circumstances, an issue examined in more detail in Chapter 7.

There is no clear distinction between CDR and SRM regarding deployment duration. Even though controlling atmospheric concentrations of GHGs by extracting them from the atmosphere would take many decades to have the desired effect of reducing global temperature, CDR would have to be continued at least until the desired target level had been reached, always recognising that that target is itself subject to review, contestation and change. Similarly, if the objective of SRM is to reduce either global or regional temperatures, it would need to be continued at least until the desired target were reached. Given that it would take longer to bring down GHG concentrations by CDR than it would to reduce GMST by SRM (Moreno-Cruz and Keith 2010), one might suppose that SRM would be required for a shorter period than CDR. However, this ignores the issue of the continuing emission of GHGs. If these are not greatly reduced, both CDR and SRM would have to be continued indefinitely; in the case of CDR to control the otherwise escalating level of anthropogenic GHG concentrations and ocean acidification, and in the case of SRM to control the otherwise escalating increase in GMST. Conversely, if GHG emissions are substantially reduced, both SRM and CDR could be phased out. These scenarios are considered in more detail in Chapter 7.

Impact time

The time lag between commencing the deployment of geoengineering and it having material global climate effect has significant intergenerational implications. The shorter the impact time, the later a decision to deploy can be left thus avoiding a premature commitment to technologies that might be considered by future generations to be less attractive than alternatives that might in due course be available to them. This could allow the deployment of SRM geoengineering to be deferred until there is an impending global warming climate catastrophe, the so-called emergency response option. This deferral of the deployment decision shifts the burden of that decision to future generations (discussed in more depth in Chapter 7). However, this places a residual responsibility on the earlier generations to 'arm the future' (Gardiner 2010) by undertaking sufficient research so that when in a future emergency geoengineering is called upon, reliable technological processes with reasonably predictable outcomes can be implemented without undue delay or incurring excessive risk. Conversely, where the impact time is long, as in the case of most CDR, an earlier generation would need to assume the responsibility and cost of commencing deployment in order that the benefits could be enjoyed by later generations.

SRM is not currently considered a high policy priority because those countries with the capability to undertake SRM do not perceive climate change to represent a sufficiently clear and present danger to warrant engaging with the international relations difficulties it provokes, or the environmental risks it may entail. CDR at climate-changing scale is equally unattractive as a policy option for current policymakers because its climatic effects will not be realised for many decades and this requires them to impose the costs of such a policy on the current generation on whose support they rely for their continued power, while the benefits will not be enjoyed until they are all, both the policymakers and their supporters, long dead.

On the other hand, if little or no action is taken to develop any of these technologies, the ability of those future generations, whose interests are the focus of so much climate change rhetoric, will be significantly circumscribed, limiting their options if emissions abatement and adaptation fail to mitigate the risks of dangerous climate change. Impact time has a material political dimension.

Permanence

The risks to future generations increase as more CO_2 is sequestered in sinks that are anything other than geologically permanent. In relation to carbon sequestration and storage (CCS) Mazzotti *et al.* report that 'the fraction retained in appropriately selected and managed geological reservoirs is very likely to exceed 99% over 100 years and is likely to exceed 99% over 1,000 years' (Mazzotti *et al.* 2005, 14). However, in a review of the empirical evidence from existing sequestration sites and extrapolations to scale these up 250 times to meet anticipated global needs from CCS, Benson and Cole (2008) are only able to leave open key questions concerning the security of these reservoirs over extended periods. Similar concerns are expressed by Azar *et al.* (2010).

The differences between these positions begins to evaporate when Mazzotti *et al.*'s criteria for 'appropriately selected and managed geological reservoirs' are examined in more detail. They presuppose institutional structures enduring for multiple generations capable of continuing to select and manage new and existing reservoirs to the same high standards implied by their proposed operational criteria. There are only a few examples of such long-lived institutions – the British Monarchy, the Icelandic Parliament, the Catholic Church, and universities not dependent on government funding (e.g. Oxford and Cambridge in the UK, Yale and Harvard in the US) – but whether they provide suitable models for continuing and infallible geoengineering governance is an open question. Nevertheless, there are also examples of such processes that have endured for decades such as the control regime for nuclear weapons and the management of nuclear waste, where what failings there have been have not resulted in any systemic widespread catastrophic consequences. From an intergenerational perspective, infallibility, or at least very high levels of operational integrity, are essential. The maintenance of the CO_2 reservoirs will be an unforgiving task because one single lapse could allow substantial, even total escape of sequestered CO_2. Some comfort might be obtained from their being multiple sites and hoping that large numbers of them did not fail as a result of a systemic failure of maintenance or design. Lackner's comments on permanence and concerns about the frailty of human system are apposite (2002).

Unless DAC and biochar installations are located above suitable geological formations, it will be necessary to create a substantial transport industry to move liquefied CO_2 and biochar products from the point of production to the point of sequestration, probably involving land/sea/land transfers. These industries would have considerable impacts on large areas of the landscape albeit that the risks of leakage from industrial accidents are unlikely to be material. To what extent the lives of future generations would be impaired by such developments is a moot point. In poorer areas, people might be happy to have the employment opportunities such a large scale industry would provide despite the local environmental impacts it might have, and the activity's essential role in combating global warming could be unimportant to them. Although, as Leach *et al.* (2012, 301) show, it does not always follow that the economic benefits produced from such activities are enjoyed by poor rural communities engaged in them. On the other hand, others might regret that earlier generations had not been more diligent in dealing with the environmental impacts of their own activities in their own time, in such a way as not to bequeath such a burdensome legacy to the future.

Residence time

Stratospheric aerosol injection (SAI) is thought to be one of the most feasible geoengineering technologies. Residence time refers to the duration during which aerosol particles remain in the atmosphere. Residence time is a function of the particle composition, the stratum of the atmosphere in which they are released, the location where they are released, and their size (Caldeira and Wood 2008; Crutzen 2006; Dickinson 1996; Izrael, Ryaboshapko, and Petrov 2009; Keith 2000; Rasch *et al.* 2008; Kravitz *et al.* 2009; Robock 2009). The current maximum residence time that is contemplated from aerosols injected into the stratosphere is of the order of two to three years making it necessary constantly to refresh them. However those injected into the troposphere would fall to Earth within days. The shorter the residence time the more frequent the sulphate injections must be in order to retain the necessary SWR reflectivity. However, the longer the residence time, and therefore the less frequent and cheaper the injections, the longer the cessation time (see below). Residence time has significant implications for the governance of geoengineering with long term ethical and operational implications for future generations. If particles could be developed that would stay aloft more or less indefinitely (Keith 2010) there would be a much greater ethical burden on the generation releasing them than if they had to be constantly refreshed. However, how that increased ethical burden is to be balanced against the benefit of lower costs that the long term residence would offer, is a question that only those that will have to confront it will be able to determine.

Termination time

Termination time refers to the time it would take to end the geoengineering activity. It is a significant intergenerational consideration that as far as I'm aware is not addressed in any detail elsewhere in the geoengineering literature. For all forms of geoengineering that entail large numbers of industrial installations or changes to agricultural practice that become embedded in local economies, there will be an inevitable reluctance to end, or even reduce them, in response to perceived negative consequences that manifest at a spatial or temporal distance from the installation or activity. Afforestation, biochar and the widespread use of reflective crops are examples of geoengineering technologies that would be difficult to stop if major global economies were created to support them. Conversely, the more concentrated the technology is in the hands of a few, the more easily it could be terminated. SAI and Earth-orbiting mirrors are two such examples.

This intergenerational difficulty does not follow the CDR/SRM categorisation. If future generations wish to phase out geoengineering whether because decarbonsing the global economy has been successfully achieved, or the negative side-effects of geoengineering have become unacceptable, their task will be made difficult if they meet resistance from large numbers of people with a vested interest in continuing the geoengineering; they will be confronted by the usual challenges of transitioning away from sunset industries – loss of livelihoods, threats to the cohesiveness of local communities and threats to the cash often generated by long-standing industries whose major capital costs have been largely amortised. This inertia parallels the current global reluctance to decarbonise the economy because the perceived fears of climate change in the distant future, or at least beyond the period of current economic planning, are less compelling drivers of behaviour than the need to maintain the quality of life today. There may be powerful intergenerational equity arguments for retaining flexibility by only deploying geoengineering technologies with relatively short termination times.

Cessation time

Cessation time refers to the inertial cooling effect once geoengineering is terminated. For most geoengineering technologies this period is sufficiently short as to not have intergenerational consequences. However others do pose certain problems. For example, if vast amounts of biochar are added to soils across the globe and it is later found that there is an imminent tipping point beyond which the enhancements to soil quality brought about by the biochar are reversed and become detrimental in some, as yet, unforeseen way, removing the excess biochar will be far from straightforward. Similarly, if aerosols with extended residence times are deployed, it could be difficult to remove them if they have significant negative unintended consequences, or future policy priorities require the deployed capacity of SRM to be reduced.

Termination problem

The final temporality of geoengineering to be considered here is that surrounding the decision to stop an SRM programme. Modelling evidence suggests that once SRM is terminated, GMST will revert in a matter of a decade or two to what it would have been had there been no such intervention (Matthews and Caldeira 2007). The termination of SRM would result in an increase in GMST at a rate dependent upon the cooling effect of geoengineering still deployed at the time of its termination. However, the amount of geoengineering activity at that point in time would itself be dependent on the level of atmospheric GHG concentration and the then actual GMST. Thus if little emissions abatement has been done with the result that GHG concentrations remain high, and it is desired to keep the GMST to no more than, say, 2°C above pre-industrial era levels, a great deal more SRM will have to be in place.

Figure 2.1 highlights the implications of the termination problem. If atmospheric GHG concentration had risen to 1,000ppmv (*x*-axis), assuming that climate sensitivity[4] is 3.5°C (*y*-axis), there would be a latent GMST increase of 9°C[5] with geoengineering being responsible for reducing that by 7°C (*z*-axis) so as to limit the increase to 2°C. On termination this 7°C cooling effect would be entirely reversed during the ensuing twenty years or so. This would be an

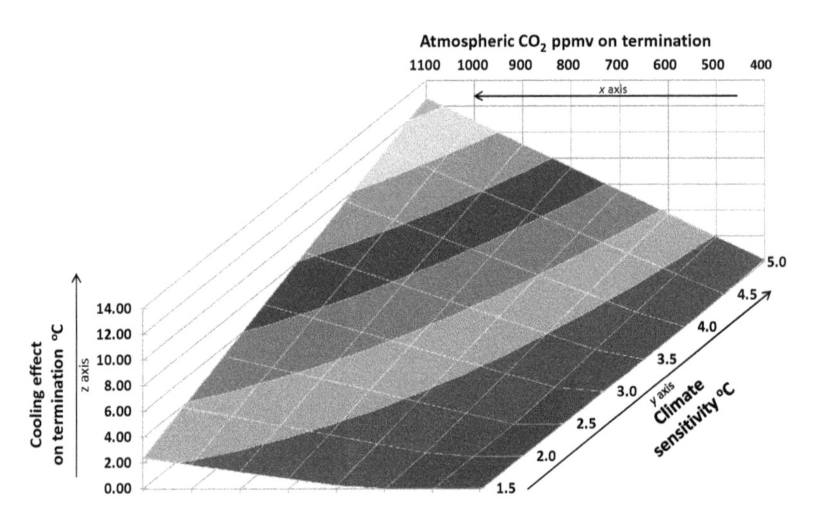

Figure 2.1 Geoengineering cooling effect at termination to limit GMST increase to 2°C above pre-industrial era

Source: compiled by the author.

unprecedented rate of increase, approaching one hundred times that over the last 200 years. There would be great concern that many species would be unable to adapt fast enough to survive, whereas a managed rise of 7°C over centennial timescales would likely be catastrophic for many fewer species. The situation would be correspondingly worse if climate sensitivity were higher than 3.5°C and vice versa (11.5°C average temperature rise with climate sensitivity at 4.5°C, and an average 2°C rise with climate sensitivity at 1.5°C).[6]

The take-away message from this graph is that the avoidance of catastrophically rapid global warming on the termination of an SRM geoengineering programme would require a low dependence on the SRM cooling effect at the time of the termination. In short, the SRM termination problem can only be mitigated by ensuring that while it is in use, there is a sufficient reduction in atmospheric GHGs so that it can be gradually scaled down and phased out over an appropriate period (Wigley 2006; Fox 2009). It also follows that even if the direct and consequential risks of geoengineering can be accommodated, there would *never* be a circumstance for it to be rational not to increase emissions abatement, for such a failure would for all practical purposes preclude the possibility of phasing out SRM geoengineering.

Summary

The spatiality and temporality of geoengineering are not easily delineated by the SRM/CDR dichotomy save for the general observation that some SRM technologies have the capacity to reduce global warming substantially more rapidly than any form of CDR. However, on most other metrics, there is a wide range of spatial and temporal variability between the technologies within each of these two categories.

Social implications

The purpose of this chapter has been to identify within the climate system the physical roots of the main theme of this book, namely the socio-political challenges presented by geoengineering as a response to climate change. The UNFCCC process already frames climate change as a global political issue, engaging 192 nation-states in its deliberations. But the threats and opportunities, responsibilities and obligations, capacities and shortcomings that climate change touches, not only arise differently between nations but also within nations such that the range of interests to be reconciled extends way beyond a mere 192. It is an anecdotal maxim of human relations that the difficulty of reaching agreement increases exponentially with the number of parties involved. This makes climate change the ultimate political challenge.

These differences arise not just from competing views about cooperation and competition in international relations, but also because the physics and chemistry of climate change affect the global population in extraordinarily diverse ways. The spatiality of climate change is already altering long-established weather patterns around which settled lifestyles have emerged and become embedded in local and regional communities. As the AOSIS example referred to earlier[3] shows, in some cases these lifestyles are already demonstrably destabilised. The threats manifest in different ways as local habitats on which human life depends are increasingly affected by climate change. The increased fickleness of the weather makes communities dependent on subsistence farming considerably more vulnerable than they have historically been in the face of inter-annual climate variability. And for those communities in developed nation-states, whose local economies are highly dependent on globalisation, climate change will increasingly test their resilience, as links between nodes in their complex global web of relations are stressed. There is nothing new in this spatial diversity, since there were human settlements, there have been challenges in sharing common resources. Climate change makes these challenges global rather than bi- or even multi-lateral.

In addition to problems of a spatial nature, climate change also presents equally intractable temporal problems. The many different temporalities described in this chapter create challenging intergenerational tensions. If the most pessimistic projections for the next few centuries come to pass, the global population will collapse to a small fraction of its current figure, a process that will entrain socio-political changes on a scale not witnessed perhaps since the Black Death wiped out a third to a half of the global population in the fourteenth century. But if something similar to this is what is meant by the UNFCCC when it refers to 'dangerous anthropogenic interference in the climate', it begs the question of how much the earlier generations are prepared to alter their behaviours to avert such a catastrophe, when those distant future generations whose interests they are being asked to protect comprise individuals they'll never know, living in places they'll never visit.

Different worldviews respond to this question in different ways (Thompson, Ellis, and Wildavsky 1990). Egalitarians think that this must require radical reductions in consumption, particularly in the affluent West; Individualists believe that technology will come to the rescue and continued economic growth is the only way to create the wealth and innovation necessary to protect those future generations from the depredations of climate change. Hierarchists believe that with due process the elite will be able to deliver a better future. While Fatalists, who form the vast majority, will make the best of whatever hand they are dealt. How are these polarised positions to be reconciled? What are the moral equivalences between the need to improve the lot of the world's poor, and the need to protect the lifestyles of the world's rich? Are these alternatives or can they both be achieved? If the poor are more vulnerable to the ravages of climate change does that imply any increased obligation by those less threatened to protect them? The Egalitarian is cautious and sceptical about the promise of progress, arguing that it is precisely yesterday's progress that has caused today's problem. The Individualist argues that technology has brought extraordinary improvement in lifestyles since the onset of the industrial era and there is no reason to suppose that this will not continue, and any difficulties that might arise along the way will be well within our capacity to confront successfully. Are the Egalitarians limiting possible responses to climate change by overstating the threats they present, or the Individualists deferring essential action by recklessly overestimating humanity's collective capacity to act in a timely and effective manner?

Resolving this question is confounded not only by the diversity of the multiple competing worldviews, but also by the irreducible uncertainty attached to projections about the future. In later chapters, the nature and policy significance of how these issues are framed will be examined in detail. For the moment, the key message from this brief survey of the various geoengineering technologies, is that it is the spatial and temporal dynamics of their physics and chemistry that create the context for the discourse at the heart of humanity's climate policy initiatives. In the next chapter I examine the conventional scientific method as the established approach to policymaking where the science itself is a central player.

Appendix: Biochar: resources to sequester $1PgCyr^{-1}$

Introduction

The purpose of this calculation is to illustrate the extent of the challenge implied in deploying biochar production as a form of geoengineering. Following Pacala and Socolow's notion of wedges, the analysis is based upon building the capacity to sequester annually 1PgC in the form of biochar within 50 years (Pacala and Socolow 2004), each wedge representing approximately 10 per cent of total anthropogenic emissions. For the reasons explained more fully below, the calculation is intended to be indicative as there are many uncertainties not only about the value of current and historical variables but particularly with regard to future innovation and political engagement with biochar. The analysis does not suggest that it would be impossible annually to sequester 1PgC in the form of biochar but it does suggest that it is most unlikely to arise from organic market forces. The primary reason for this is that the volumes of material (biomass feedstock, biochar and waste) implicit in a $1PgCyr^{-1}$ programme are orders of magnitude greater than any current commercial activity in this, or indeed in most other industrial sectors. In addition, the environmental implications of scaling biochar to this level are untested whether in terms of sourcing the quantity of raw material necessary, moving it through the various stages of the process, or sequestering it in terrestrial or marine sinks. Finally, scaling other CDR technologies to deliver more sequestration wedges of a similar size would compound these challenges by virtue of incremental and competing demands on some of the same resources (e.g. land, labour, freshwater, and transport).

Biochar – the background

Biochar is defined as charcoal manufactured with the express purpose of enriching soils (Meyer, Glaser, and Quicker 2011, 9473). There are several different technologies for its manufacture which produce charcoal with a wide range of physical characteristics, some are better suited to specific applications for example as a catalyst in steel production and for medical uses. According to the UN Food and Agriculture Organisation database, world manufacture of charcoal was 49Tg in 2011. It has followed a power growth trend since the FAO records began in 1961 and if the same trend continued, global annual production would reach 150Tg within 50 years. By comparison world coal production was five orders of magnitude greater at 7.7Pg (2011 data from the World Coal Association).[7]

Biochar is thought to enrich certain types of soil but there is limited research on the global capacity for biochar as a soil enhancer or whether there are any negative effects from continuous applications or if there are saturation thresholds.

Discussion

The following calculation provides an indicative assessment of the land, transport and manufacturing facilities required to produce $1PgCyr^{-1}$ in the form of biochar. There are other significant questions relating to biochar as a means of geoengineering not addressed here, for example the extent to which any carbon sequestered by biochar is offset by emissions from its production (through the entire life cycle of producing the biomass, harvesting it, transporting it to biochar production facilities for conversion to charcoal and then distributing it for incorporation in the soil). Where biomass is grown specifically for biochar all the emissions in every stage need to be accounted for. However, where the biomass is a waste product from some other agricultural or industrial process, some of the emissions may more appropriately be allocated to the primary activity from which the biomass waste is produced. It is because these factors will be difficult to quantify in the absence of detailed research that they are not considered further here. This calculation focuses on quantities that can be estimated from available data.

Several variables in the calculation are subject to a wide range of possible values which are explained in the table below. Low, medium and high values are given for those variables. It is outside the scope of this brief analysis to examine the dynamics and interdependencies of these variables, rather the calculation is an attempt to assess some broad parameters in order to better understand the scale of the resource requirement if biochar were to be adopted as a significant contribution to sequestering atmospheric carbon.

Input data has been obtained from peer reviewed journal papers and from the World Bank databank and the UK Forestry Commission. Sources are shown for each datum.

The calculation indicates that the land requirement is in the range of 1.9Mkm^2 to 25Mkm^2 with a medium figure of 4.2Mkm^2. By comparison the European Union covers 4.3Mkm^2, USA 9.8Mkm^2 and Russia 16.4Mkm^2, and current global forests cover 40Mkm^2 and all agricultural land covers 50Mkm^2. The calculated areas assume that all feedstock is timber grown especially for biochar. The use of waste products from other industrial processes would reduce this land requirement. The substitution of other forms of biomass that sequester carbon more slowly may increase it. There are also wide regional differences in the rate at which carbon is sequestered by biomass.

The number of biochar production units is in the range of 11,000 to 100,000. The medium estimate (33,000) compares favourably with an estimate of 45,000 installations calculated from data provided to the author by Prof. Bridgwater.[8] The lower estimate implies considerably higher productivity than is achieved in the current generation of charcoal manufacturing plants. Assuming that on average current plants have an output of 10Mgyr^{-1}, they number something of the order of 5,000. This figure compares with the requirement of 100,000 that assumes continued relatively low productivity. However, the medium figure requires 13 plants, each producing 50Tg of biochar per year, to be built every week for 50 years to produce 1PgCyr^{-1} in the form of biochar.

The biomass feedstock requirement each year would lie in the range 2–12.5Pg of dry wood with the medium figure being about 3.5Pg (by comparison global annual coal production is currently 7.7Pg). In volume this corresponds to between 100 and 1,000 million TEUs[9] with the medium figure being about 200 million TEUs.[10] The biochar produced would have a volume of between 40 and 200 million TEUs with the medium figure of about 150 million TEUs. In aggregate this amounts to an annual transport requirement of between 140 and 1,200 million TEUs. By comparison, current world port container shipments amount to 602 million TEUs with each container being used on average 18 times per year. On this basis, biochar production would require between 7 and 70 million (medium figure 20 million) dedicated 20-foot containers (or equivalent) to transport feedstock to the production facilities and biochar away from them. This computation ignores the removal of waste products which would add substantially to this number given that the mass of feedstock could be as much as five times the biochar output (Bridgwater suggests three times). In addition, many of the waste products are environmentally unfriendly and would require special disposal arrangements if produced at this scale, although some will be marketable, for example, as low grade fuel (Bridgwater).

Table A2.1 Biochar computation

	Value			*Units*	*Source*	*Comments*
	Low	Medium	High			
INPUT DATA						
Carbon yield g per ha per year	2.00x1	4.00×10^6	6.00×10^6	g	a.	A yield of $4 MgCha^{-1} yr^{-1}$ quoted by UK Forestry Commission as being slightly higher than yields empirically measured in UK.
Density of wood	330	420	560	kgm^{-3}	b.	Different wood species have a wide range of densities. Other forms of biomass (e.g. corn stover, palm kernel shells etc.) have significantly different densities.
Carbon content of wood	40%	50%	60%		c.	Most dry plant material contains about 50% by weight of carbon.
Density of biochar	250	300	750	kgm^{-3}	d.	Biochar density varies considerably according to the feedstock and the processing method. High densities are most often found in specialist applications e.g. steel production and medical applications.
Carbon content of biochar	51%	60%	90%	by wt	e.	Carbon content of biochar varies considerably according to the processing method. For high volume applications the highest densities are probably uneconomic as they require longer retention time in the kilns.

	Value			*Units*	*Source*	*Comments*
	Low	Medium	High			
Carbon yield	20%	60%	88%	by wt	f.	The ratio of carbon in biochar product to carbon in wood feedstock varies considerably depending upon feedstock and processing method.
Global land area	129,710,339	129,710,339	129,710,339	km^2	g.	Global land area excluding inland water bodies
Global agricultural land	48,827,130	48,827,130	48,827,130	km^2	h.	Land that is arable, under permanent crops, and under permanent pastures
Global forest	40,204,180	40,204,180	40,204,180	km^2	k.	Land under natural or planted stands of trees of at least 5 meters in situ
1 TEU (twenty foot unit)	38.5	38.5	38.5	m^3		Standard container unit used in transportation
2012 TEU port shipments	6.02×10^8	6.02×10^8	6.02×10^8		l.	
Movements per annum per container	18	18	18		m.	
Biochar annual production per plant	2.00×10^{10}	5.00×10^{10}	1.00×10^{11}	g	n.	

Table A2.1 (continued)

	Value			Units	Source	Comments
	Low	Medium	High			
CALCULATIONS FOR 1PgC IN FORM OF BIOCHAR						
Amount of biochar						
Gross weight of biochar containing 1PgC	1.96×10^{15}	1.67×10^{15}	1.11×10^{15}	g		
Volume of that biochar	7.84×10^{9}	5.56×10^{9}	1.48×10^{9}	m^3		
Volume in TEUs	2.04×10^{8}	1.44×10^{8}	3.85×10^{7}	TEUs		
To convert C_{wood} to $C_{biochar}$						
Wood C required to make 1PgC in biochar	5.00×10^{15}	1.67×10^{15}	1.14×10^{15}	g		
Gross weight of dry wood containing this amount of C	1.25×10^{16}	3.33×10^{15}	1.89×10^{15}	g		
Volume of wood	3.79×10^{10}	7.94×10^{9}	3.38×10^{9}	m^3		
Volume of wood in TEUs	9.84×10^{8}	2.06×10^{8}	8.78×10^{7}	TEUs		
Land requirement						
Land required to yield 1PgC in form of biochar	2.50×10^{7}	4.17×10^{6}	1.89×10^{6}	km^2		
% of total land surface not already forested	27.9%	4.7%	2.1%			
% of total agricultural	51.2%	8.5%	3.9%			

	Value			*Units*	*Source*	*Comments*
	Low	Medium	High			
Container requirement	$6.60x10^7$	$1.95x10^7$	$7.02x10^6$			
% of 2010 container traffic	221%	65%	23%			
Number of biochar plants	98,039	33,333	11,111			
Plants built per week for n years						
20	94	32	11			
50	38	13	4			
100	19	6	2			

Sources:
a (Broadmeadow and Matthews 2003, Figure 3 caption).
b (Broadmeadow and Matthews 2003, 8).
c (Broadmeadow and Matthews 2003, 1).
d (Pastor-Villegas *et al*. 2006, 106).
e (Meyer, Glaser, and Quicker 2011, Table 2).
f (Meyer, Glaser, and Quicker 2011, Table 2).
g World Bank.
h World Bank.
k World Bank.
l World Bank.
m World Bank.
n No reliable data available. Estimates based on current and proposed plants of 10,000 to 30,000 $Mgyr^{-1}$ and arbitrary assumption about possible expansion of production facilities to a maximum of 100,000 $Mgyr^{-1}$.

Notes

1 The Royal Society (Shepherd *et al.* 2009) comparison table of the relative merits of different geoengineering technologies shows all SRM options, except painting urban roofs white, as being less safe than all CDR options, except ocean fertilisation. They also suggest that the ethical issues raised by CDR are less severe than those provoked by SRM.
2 The virtual absence of empirical field research into any form of geoengineering has continued even since Blackstock, Victor and their associates were writing in 2009.
3 See for example, the 2012 declaration of the Alliance of Small Island States (AOSIS), available online at http://aosis.org/wp-content/uploads/2012/10/2012-AOSIS-Leaders-Declaration.pdf (accessed 4 February 2015).
4 Climate sensitivity is defined as the change in global mean surface temperature that is caused by a doubling of the atmospheric CO_2 concentration.
5 Calculated as follows (1000–280)/280 x 3.5 = 9 where 280 is the atmospheric CO_2 concentration in ppmv at the beginning of the industrial era and climate sensitivity is constant at 3.5°C. It is assumed that climate sensitivity is a constant but the possibility that it may vary in response to changes in other climate variables cannot be ignored and it is therefore possible that GMST may rise faster or slower than the assumed constant value implies. It is also assumed that regional surface temperature changes will vary in proportion to the global mean. This assumption is almost certainly invalid which raises the prospect of spatial variations in which different regions fare disproportionately better or worse at different values of climate sensitivity.
6 IPCC AR5 estimates climate sensitivity to be likely to be in the range of 1.5°C to 4.5°C (IPCC 2014, SPM D. 2).
7 www.worldcoal.org/resources/coal-statistics/.
8 Tony Bridgwater is a Professor of Chemical Engineering at Aston University in Birmingham, UK, specialising in thermal conversion of biomass for production of fuels and chemicals.
9 TEU stands for *twenty foot unit*. A TEU is the standard size of a container used for international shipping. It is $38.5m^3$ in volume.
10 These volume calculations assume that there is no shrinkage between the dry and green timber. Any shrinkage would increase the volume of biomass to be harvested and transported.

References

Adger, W. N. 2003. 'Social Capital, Collective Action, and Adaptation to Climate Change'. *Economic Geography* 79(4): 387–404.

Arrow, Kenneth J. 1963. 'Uncertainty and the Welfare Economics of Medical Care'. *American Economic Review* 53(5): 941.

Auffhammer, Maximilian, V. Ramanathan, and Jeffrey R. Vincent. 2012. 'Climate Change, the Monsoon, and Rice Yield in India'. *Climatic Change* 111(2): 411–424. doi:10.1007/s10584–10011–0208–0204.

Azar, Christian, Kristian Lindgren, Michael Obersteiner, Keywan Riahi, D. P. van Vuuren, K. den Elzen, Kenneth Möllersten, and Eric Larson. 2010. 'The Feasibility of Low CO2 Concentration Targets and the Role of Bio-Energy with Carbon Capture and Storage (BECCS)'. *Climatic Change* 100(1): 195–202. doi:10.1007/s10584–10010–9832–9837.

Benson, Sally M., and David R. Cole. 2008. 'CO2 Sequestration in Deep Sedimentary Formations'. *Elements* 4(5): 325–331. doi:10.2113/gselements.4.5.325.

Bickel, J. Eric, and Lee Lane. 2009. `An Analysis of Climate Engineering as a Response to Climate Change'. Copenhagen Consensus on Climate. Copenhagen Consensus Center.

Blackstock, Jason, D. S. Battisti, Ken Caldeira, D. M. Eardley, J. I. Katz, David W. Keith, A. A. N. Patrinos, D. P. Schrag, R. H. Socolow, and S. E. Koonin. 2009. 'Climate Engineering Responses to Climate Emergencies'. *Novim*, July. http://arxiv.org/abs/0907.5140.

Broadmeadow, Mark, and Robert Matthews. 2003. 'Forests, Carbon and Climate Change: The UK Contribution'. *Forestry Commission Information Note* 48(2).

Caldeira, Ken. 2009. *Geoengineering: Assessing the Implications of Large-Scale Climate Intervention.* US House of Representatives Committee on Science and Technology, Washington, DC. http://democrats.science.house.gov/Media/file/Commdocs/hearings/2009/Full/5nov/Caldeira_Testimony.pdf.

Caldeira, Ken, and Lowell Wood. 2008. 'Global and Arctic Climate Engineering: Numerical Model Studies'. *Philosophical Transactions of the Royal Society A: Mathematical, Physical and Engineering Sciences* 366(1882): 4039–4056. doi:10.1098/rsta.2008.0132.

Crutzen, Paul J. 2006. 'Albedo Enhancement by Stratospheric Sulfur Injections: A Contribution to Resolve a Policy Dilemma?'. *Climatic Change* 77(3–4): 211–220. doi:10.1007/s10584–10006–9101-y.

Dickinson, Robert E. 1996. 'Climate Engineering a Review of Aerosol Approaches to Changing the Global Energy Balance'. *Climatic Change* 33(3): 279–290. doi:10.1007/BF00142576.

Fox, Tim. 2009. *Climate Change: Have We Lost the Battle?*. Institution of Mechanical Engineers. www.imeche.org/NR/rdonlyres/77CDE5E4-CE41-4F2C-A706-A630569EE486/0/IMechE_MAG_Report.PDF.

Gardiner, Stephen. 2010. 'Is "Arming the Future" with Geoengineering Really the Lesser Evil?'. In *Climate Ethics: Essential Readings*, edited by Stephen Gardiner, Simon Caney, Dale Jamieson, and Henry Shue, 285–312. OUP USA.

Govindasamy, B., and Ken Caldeira. 2000. 'Geoengineering Earth's Radiation Balance to Mitigate CO2 Induced Climate Change'. *Geophysical Research Letters* 27(14): 2141–2144.

Hulme, Mike. 2010. 'Cosmopolitan Climates'. *Theory, Culture & Society* 27(2–3): 267.

IPCC. 2014. *Climate Change 2013 – The Physical Science Basis: Working Group I Contribution to the Fifth Assessment Report of the Intergovernmental Panel on Climate Change.* Cambridge University Press.

Izrael, Yu., A. Ryaboshapko, and N. Petrov. 2009. 'Comparative Analysis of Geo-Engineering Approaches to Climate Stabilization'. *Russian Meteorology and Hydrology* 34(6): 335–347. doi:10.3103/S1068373909060016.

Jasanoff, Sheila. 2010. 'A New Climate for Society'. *Theory, Culture & Society* 27(2–3): 233–253.

Keith, David W. 2000. 'Geoengineering the Climate: History and Prospect'. *Annual Review of Energy & the Environment* 25(1): 245.

Keith, David W. 2010. 'Photophoretic Levitation of Engineered Aerosols for Geoengineering'. *Proceedings of the National Academy of Sciences* 107(38): 16428–16431. doi:10.1073/pnas.1009519107.

Kravitz, Ben, Alan Robock, Luke Oman, G. Stenchikov, and A. B. Marquardt. 2009. 'Sulfuric Acid Deposition from Stratospheric Geoengineering with Sulfate Aerosols'. *Journal of Geophysical Research* 114.

Lackner, K. S. 2002. 'Carbonate Chemistry for Sequestering Fossil Carbon'. *Annual Review of Energy and the Environment* 27(1): 193–232.

Leach, Melissa, James Fairhead, and James Fraser. 2012. 'Green Grabs and Biochar: Revaluing African Soils and Farming in the New Carbon Economy'. *Journal of Peasant Studies* 39(2): 285–307.

Lempert, R. J., and D. Prosnitz. 2011. *Governing Geoengineering Research: A Political and Technical Vulnerability Analysis of Potential Near-Term Options*. RAND Corporation.

Matthews, H. D., and Ken Caldeira. 2007. 'Transient Climate–carbon Simulations of Planetary Geoengineering'. *Proceedings of the National Academy of Sciences* 104(24): 9949–9954. doi:10.1073/pnas.0700419104.

Mazzotti, M., J. C. Abanades, R. Allam, K. S. Lackner, F. Meunier, E. Rubin, J. C. Sanchez, K. Yogo, and R. Zevenhoven. 2005. 'Mineral Carbonation and Industrial Uses of Carbon Dioxide.' *IPCC Special Report on Carbon Dioxide Capture and Storage*, 321–338.

Meyer, Sebastian, Bruno Glaser, and Peter Quicker. 2011. 'Technical, Economical, and Climate-Related Aspects of Biochar Production Technologies: A Literature Review'. *Environmental Science & Technology* 45(22): 9473–9483. doi:10.1021/es201792c.

Mitchell, David L., and William Finnegan. 2009. 'Modification of Cirrus Clouds to Reduce Global Warming'. *Environmental Research Letters* 4(4): 045102.

Moreno-Cruz, J. B., and David W. Keith. 2010. *Climate Policy under Uncertainty: A Case for Geo-Engineering*. http://works.bepress.com/cgi/viewcontent.cgi?article=1000&context=morenocruz&sei-redir=1#search=%22moreno+cruz+keith+uncertainty%22.

Pacala, S., and R. Socolow. 2004. 'Stabilization Wedges: Solving the Climate Problem for the Next 50 Years with Current Technologies'. *Science* 305(5686): 968–972. doi:10.2307/3837555.

Pastor-Villegas, J., J. F. Pastor-Valle, J. M. Rodríguez, and M. García García. 2006. 'Study of Commercial Wood Charcoals for the Preparation of Carbon Adsorbents'. *Journal of Analytical and Applied Pyrolysis* 76(1): 103–108.

Rasch, Philip J., Simone Tilmes, Richard P. Turco, Alan Robock, Luke Oman, Chih-Chieh (Jack) Chen, Georgiy L. Stenchikov, and Rolando R. Garcia. 2008. 'An Overview of Geoengineering of Climate Using Stratospheric Sulphate Aerosols'. *Philosophical Transactions of the Royal Society A: Mathematical, Physical and Engineering Sciences* 366(1882): 4007–4037. doi:10.1098/rsta.2008.0131.

Rayner, Steve. 2010. 'The Geoengineering Paradox'. *The Geoengineering Quarterly*, March.

Robock, Alan. 2009. *Geoengineering: Assessing the Implications of Large-Scale Climate Intervention*. Washington, DC. http://democrats.science.house.gov/Media/file/Commdocs/hearings/2009/Full/5nov/Robock_Testimony.pdf.

Schneider, Stephen H. 2008. 'Geoengineering: Could We or Should We Make It Work?'. *Philosophical Transactions of the Royal Society A: Mathematical, Physical and Engineering Sciences* 366(1882): 3843–3862. doi:10.1098/rsta.2008.0145.

Shepherd, John, Ken Caldeira, P. Cox, J. Haigh, David W. Keith, B. Launder, Georgina Mace, G. MacKerron, J. Pyle, and Steve Rayner. 2009. 'Geoengineering the Climate: Science, Governance and Uncertainty'. The Royal Society.

Thompson, Michael, Richard Ellis, and Aaron Wildavsky. 1990. *Cultural Theory*. Westview Press.

Tilmes, Simone, Rolando R. Garcia, Douglas E. Kinnison, Andrew Gettelman, and Philip J. Rasch. 2009. 'Impact of Geoengineered Aerosols on the Troposphere and Stratosphere'. *Journal of Geophysical Research: Atmospheres* 114 (D12): D12305. doi:10.1029/2008JD011420.

Vaughan, N. E., and T. M. Lenton. 2011. 'A Review of Climate Geoengineering Proposals'. *Climatic Change* 109(3): 745–790. doi:10.1007/s10584-10011-0027-0027.

Victor, David G., M. Granger Morgan, Jay Apt, John Steinbruner, and Katharine Rich. 2009. 'The Geoengineering Option'. *Foreign Affairs* 88(2): 64–76.

Wigley, T. M. L. 2006. 'A Combined Mitigation/Geoengineering Approach to Climate Stabilization'. *Science* 314(5798): 452–454. doi:10.1126/science.1131728.

3 The limits of reductionism

> *Critics see geoengineering as just the latest chapter in a long story of command and control, which includes Francis Bacon's sordid attempts to wrest the secrets out of Mother Nature …*
>
> (Buck 2012, 257)

> *[Many considering the implications of Bacon's ethics] ask whether we should control nature rather than whether we can ultimately control nature or know the limits of our control. One possible result of this oversight is that Baconian attitudes and epistemological assumptions continue to figure more or less prominently in national and globally influential institutions such as the UK's Royal Society, the US's National Science Board, IPCC and World Commission on Environment and Development.*
>
> (Charlesworth and Okereke 2010, 125–126)

Only for the likes of Dr Frankenstein does reverse engineering a complex life form from spare body parts not remain an elusive skill. Is it just a matter of time before, in the real world, we acquire the necessary knowledge and technology to build a new life that is not a Monster, or is there something structural that renders this forever a hopeless aspiration? In Baconian terms, geoengineering asks much the same questions of the climate system. The challenge inherent in the claim that geoengineering has a role to play in mitigating the effects of climate change is that we can decide upon a generally acceptable global climate, and then create it by the application of science and technology without having to sacrifice much of what we consider makes our lives worthwhile. This chapter examines our conventional approach to that challenge.

Different frames produce different, possibly even conflicting, ways of approaching the underlying questions. Is narcotics primarily a matter of criminal supply or public demand? Is education primarily about realising each individual's potential, or is it to produce the right mix of talents to drive the national economy? Clearly these '*a*' or '*b*' framings are gross oversimplifications, there may be '*c*'s' and '*d*'s' or more, and they may be linked by '*ands*' not '*ors*'. Fischer (2003, 144) explains that:

> Frames […] determine what the actors will consider the 'facts' to be and how these lead to normative prescriptions for action. Moreover, one

> cannot simply compare different perspectives for dealing with a problem without recognizing that frames change the problem.
>
> (Fischer 2003, 144)

Given their different framings, today's dominant triple of ideologies, liberalism, socialism and conservatism (Alexander 2014) select different facts and apply different normative criteria to most issues of significant social concern, and in particular to those that engage with equality and fairness. Applying Alexander's notion of a 'fundamental criterion' (the standard by which an ideology exhorts its adherents 'to do this rather than that'), to assessing the challenges provoked by climate change, we might surmise that for the liberal, climate change represents an existential threat to individual agency; for the socialist it is an existential threat to society; whilst for the conservative, if it is a threat at all, and that is far from clear, it is a threat to the integrity of what humanity has accumulated and will pass to the future. Framing matters.

The importance of framing is illustrated by the US Department of Defense report (*Climate Change Adaptation Roadmap, 2014 2014*). Recognising climate change as a global phenomenon and a security threat to US interests they preface their analysis by stating that:

> Politics or ideology must not get in the way of sound planning. Our armed forces must prepare for a future with a wide spectrum of possible threats, weighing risks and probabilities to ensure that we will continue to keep our country secure.

For those in the US, who are the primary audience for this document, it may be understood as challenging the ideologically driven position taken by many conservatives. However, taken at face value, the suggestion that the US military plans without allowing politics or ideology 'to get in the way' would suggest that for them the 'American way of life' that they are responsible for protecting is somehow elevated beyond politics and ideology. Many, if not most, outside the US might consider this framing of climate change to be extremely provocative given the historical role of US emissions in the course of climate change.

Climate change as a problem

It is in the nature of problems that they have solutions, however elusive they may be. One need only rise to the challenge and Bacon's aspiration of humanity's dominion over nature can be realised. Common to all three leading ideologies' dominant framing of climate change, is the notion that if it is a problem, then either accumulating anthropogenic carbon emissions are its cause, or if not, they greatly exacerbate it. Buoyed by the success of the 1987 Montreal Protocol in phasing out emissions that depleted stratospheric ozone, the international community approached the Rio Summit in

1992 initially imagining that reducing GHG emissions would be much like reducing ozone-depleting emissions. After a few years, as the political feasibility of this became increasingly uncertain, the idea of adaptation was introduced so as to be able to cope better with whatever ravages climate change may inflict upon us. And within the last decade, as concerns have begun to surface that neither emissions abatement nor adaptation will be either timely enough or at sufficient scale to avert the feared dangerous climate change, the idea of intervening directly in the global climate to head off the problem has risen rapidly up the agenda. The framing throughout has been that climate change is a problem and humanity's task is to solve it; our tools are emissions abatement, adaptation and geoengineering.

In the quotation at the head of this chapter, Buck captures the views of many when she refers caustically to Bacon's desire to dominate nature, to exploit it for the instrumental benefit of humanity. She is, perhaps, being a little harsh since Bacon's late sixteenth and early seventeenth century reality was very different from today's. London had a population of about 250,000 and even a century later Manhattan's population was little more than 5,000. In a world before Darwin's theory of evolution, before Faraday's electric motor and Watt's steam engine, and Mendeleev's Periodic Table, and relativity and quantum mechanics, and even before Harrison had solved the longitude problem with his precision engineered clocks enabling us to chart a route across a featureless ocean, understandings of the natural world that for us are commonplaces, were then still shrouded in mystery and superstition. We too easily forget that even the great Sir Isaac Newton, whose influence on the scientific era has been immense and enduring, was also a practising alchemist. Bacon should be forgiven his hyperbolic ambitions in his quest to clear away the thick veil of ignorance that shrouded his world. With the limited tools and only animal, wind and water power at his disposal, I conjecture that while Bacon might have been enticed by the notion of the Anthropocene, he would have dismissed it as a very distant prospect.

Yet, little more than four centuries later, the concept of the Anthropocene captures today's world. Humanity now has the demonstrable capacity to dominate nature both wittingly and unwittingly. Whether it also has the capacity to bend nature to its will is a different question addressed by Charlesworth and Okereke (2010). They observe that the focus by Buck and others on the ethics of Bacon's project to subordinate nature to human desires, fails to give sufficient weight to the equally important and prior question as to whether it is feasible. If it is not, then its ethics become a mere intellectual game of no practical consequence. They maintain that it is 'a large leap of faith to assume that humans can successfully control, manage or know the "tensile strength" of the Earth System as [the] scientific intelligentsia would seem to believe" (2010, 126).

In reflecting on this question it is important not to conflate the local and the global. Humanity demonstrably has the power to dominate nature at local scale – agriculture, animal husbandry, and forestry are just a few examples of such interventions. But there are global systems such as the climate, and the carbon, nitrogen and hydrologic cycles where the complexities of their interactions are such that literally taking control of these is a challenge of a different order, and one whose feasibility is not yet established.

We are children of the Enlightenment. We approach challenges by reducing them to their components parts, and using empirical observations as a basis to induce a general theory that allows the construction of a viable causal chain linking inputs to outputs. From here it is but a short step, at least in principle, to devise an intervention that will perturb the system sufficiently to produce a predictable,[1] and one hopes, more desirable outcome than would have occurred without the intervention, but not so much as to produce outcomes even more undesirable than the ones we hope to protect against. In policymaking terms this is the 'predict and act' approach. In the broader context of knowledge acquisition, Descartes and Bacon formalised this practice, setting us firmly on the way out of the mediaeval mindset, towards the industrial era and beyond. The Enlightenment has served us well but is it *The End of Reasoning*, as Fukuyama (1989) might have put it, or perhaps just another paradigm whose boundaries are being tested as impudent reality refuses to conform to rules imposed on it? This chapter argues that responding to climate change is one of many challenges that defy Descartes and Bacon's Enlightenment project.

The Enlightenment Project

Reductionism has a long history dating back at least to Democritus (c400 BCE) and his materialist account of the universe, but it became the cornerstone of modern science and technology with the onset of the Enlightenment and in particular the philosophies of Descartes and Bacon. In his 1637 treatise *Discourse on the Method for Reasoning Well and for Seeking Truth in the Sciences*, Descartes set out four rules to govern his philosophical enquiries, the second of which is 'to divide each difficulty [...] into as many parts as possible and as might be necessary to resolve it better'. The central assumption of this reductionist approach is that by an examination of the parts it is possible to understand the whole. Born a few decades before Descartes, English philosopher-statesman Sir Francis Bacon is generally regarded as having introduced induction into Enlightenment thinking, a process that enables generalised principles to be derived from specific observations.

Much of the progress made in the natural sciences and technology since the seventeenth century, progress that has been at the heart of the extraordinary growth in population and human well-being over the last 400 years, rests on the application to empirical evidence of Descartes' reductionism and Bacon's inductive reasoning. The so-called Laws of Nature all rest on generalisations from a vast enterprise of observation and induction. But induction cannot provide certainty, and as new more precise observations unearth anomalies that do not fit the established laws, so these have been refined, and in some cases abandoned, in a continuous iterative process of improvement that has come to be known as the 'scientific method'. Thus Newtonian physics gave way to the revelations of Einstein's relativity; phlogiston yielded to oxygen, and Steady State succumbed to the Big Bang theory of the universe. All theories of the natural world, a world that includes humanity, are contingent and subject to amendment in the light of continuously unfolding knowledge.

This reductionist approach to understanding the world about us reached its apogee in the logical positivism of the Vienna Circle, a collaboration of intellectuals in the inter-war years of the twentieth century whose ideas spread to the USA and UK as its members fled Nazi persecution. The central tenet of logical positivism is that reality is independent of our observations or knowledge of it (Durand and Vaara 2006). It follows that only knowledge that is empirically validated can 'count as true knowledge'. Furthermore, it holds that the universe comprises discrete entities that can be arranged hierarchically, that these entities and events can be linked causally, and most importantly, that the universe is 'completely determined and totally predictable' (Mörçol 2001, 107).

Reductionism has become so deeply embedded in the Western scientific tradition that it has infiltrated all aspects of the knowledge process. Brigandt and Love (2008) identify three interrelated kinds of reductionism – epistemological, ontological and methodological. They define epistemological reductionism as 'the idea that the knowledge about one scientific domain can be reduced to another body of scientific knowledge', for example that all biology can be explained in terms of physics and chemistry. Ontological reductionism holds that everything is physical and is constituted of nothing but its constituent molecules and their interactions. Finally, methodological reductionism argues that the functioning of systems comprising many parts can be accounted for by an examination of their parts. Our focus is primarily on this last sense that, in the context of climate change, implies that by analysing the many variables that impact the climate we can predict future climate states. By extension, in order to avert dangerous climate change, methodological reductionism allows us to model a range of interventions and predict their effects in order to then select that which most conforms to some exogenous normative criteria for desirability.

Reductionism and induction are well suited to resolving problems in closed deterministic systems. The researcher's task is to apply his skill and imagination to constructing an hypothesis about how the relevant entities and events are linked in a causal chain, and then to design experiments in order to generate sufficient observations to corroborate or amend it. This is an iterative process that is repeated as often as necessary in order to arrive at a point where the problem can be declared 'solved'. The nature of the solution is that the original problem's causal chain can be manipulated in order to arrive at alternative more desirable outcomes when, in the future, that need arises. It is in the nature of these iterations within a closed and deterministic system that the experiment is repeated in controlled conditions, always returning to the same well-defined set of starting conditions. Scientists build on the already accumulated wall of knowledge, imbricating their own new insights, in a continuous process of discovery. In this way new theories and devices are invented and existing ones improved, new materials developed with novel properties, and new ways of combining them to perform some function thought by someone to be worth performing.

This is the stuff of Kuhn's 'normal science'. In his seminal book *The Structure of Scientific Revolutions* (1962), he explains that normal science is conducted within a mindset that assumes a broad understanding of how nature works; the scientist's task is, as it were, to join the dots with the prior expectation of the kind of picture they'll create. The purpose of normal science is to resolve anomalies within the current worldview, not to unseat it. He refers to this constrained vision as a 'paradigm'. The Enlightenment paradigm is one whose normality is a deterministic universe and whose secrets will yield to reductive and inductive processes of enquiry that reveal robust causal chains linking entities and events in the past with predictable entities and events in the future.

Much happens in closed deterministic systems. Shooting balls round a billiard table; designing and constructing all manner of modes of transport from rollerblades to rockets; antibiotics that target specific bacteria; our world is replete with systems and processes with reliably predictable outcomes. Indeed, the essence of the industrial era has been the proliferation of physical, chemical and biotic creations to enhance humans' ability to eat, to travel, to communicate, to cure, to relax, to learn, to defend, to build and generally to do more, faster, safer, or cheaper, than would have been possible without them. We are consummate and innovative problem solvers. Faced with a particularly difficult problem, we build models, historically physical ones but today they are as likely to be virtual, that enable us to experiment, to test and to fail safely as we edge our way towards a version considered to be sufficiently viable to be declared a solution And then the most creative move on to the next problem, while lesser mortals make marginal improvements to existing solutions. This is the Enlightenment paradigm in action.

Not all problems can be so easily defined and in many cases it is unclear what would constitute a solution. For example, creating a mobile telephone is a well defined problem. It might require a vast array of technologies to be brought together, but if the device fits in your pocket and you can call anyone from anywhere, then the problem is solved. Conversely, by what criteria do we define the problems of juvenile delinquency, obesity, or drug addiction, and by what other criteria might we declare them as solved? The prevailing paradigm might be sufficient to build a detention or a treatment centre, but not so adept at determining what goes on inside those buildings or how to identify those requiring treatment and get them to commit to the rehabilitation programmes. While there might be little disagreement between liberals, socialists and conservatives about whether the technical problem of wireless telephony is solved, they are likely to hold deeply contested views about the nature and causes of, and remedial approaches to delinquency, obesity and drug addiction. Where these differences are ideological they are rarely capable of resolution on the basis of empirical observation, reduction and induction. The Enlightenment paradigm falls short.

It is in the nature of paradigms that they are ephemeral. If we no longer see the world about us as a closed deterministic system comprising entities and events that are causally linked and predictable, then the Enlightenment paradigm must give way to something new. But something as deeply rooted as a paradigm requires time and resolutely irresolvable anomalies before it is replaced by a successor. Kuhn explains how conservative forces resist acknowledging that their dearly held worldview has been found materially wanting. Those that come to terms with a new worldview that does account for the previously anomalous observations, participate in the future, those that don't 'are simply read out of the profession, that thereafter ignores their work'. However, while there might have been a eureka moment when the new paradigm crystallised in some genius' mind, it may be a long time before it becomes accepted as such by the wider community. The transition beyond reductionism is currently in that in-between pupal state.

A modern example of this process at work is the debate over homeopathy, a system of medicine that uses ultra-high dilutions of a wide range of plant and mineral substances.[2] Homeopathy has been practised for over 200 years during which time it has had a turbulent relationship with orthodox medicine. Today, despite many conventionally qualified doctors being practitioners of homeopathy, it is routinely denounced as being without any scientific justification. This argument is fought asymmetrically, with those for homeopathy acknowledging that they don't know how it works but protesting that they have vast stores of anecdotal evidence that it does. While those against argue that anecdotal evidence isn't scientific and moreover, for there to be an effect there must be a cause and no effect can be caused by a remedy that has no active material agent. Therefore homeopathy is rubbish – QED. Often, and probably unwisely, the homeopaths, wanting to beat their opponents at their own game using traditional reductionist analytical processes, mount defences using statistical techniques applied to patient outcomes. But these are nearly always flawed by inappropriate methodologies deriving from a forlorn attempt to overcome a fundamental and irreconcilable distinction between the two systems; while homeopathy treats patients holistically, orthodox medicine treats them reductively. If this dispute were ever resolved in favour of homeopathy it would signal a new paradigm in which the logical positivism of medical orthodoxy, that treats the human body as little more than the sum of its parts, has been displaced from its current hegemony.

Empirical data is vital to reductionism in order to induce the robust causal chains that it presupposes link entities and events. Causal chains not induced from sound empirical evidence amount to little more than speculation. It follows that for the Enlightenment paradigm to provide reliable predictions, it must be serviced with data that are sufficient, appropriate and reliable in order to sustain the process of induction. Where the predictions are interpolations between events that are either past or repeatable, it is usually not too challenging to test their reliability, even if on occasions it might be necessary to build extraordinary experimental devices such as the Large Hadron Collider[3] to do this. But where they are extrapolations into the future, their reliability cannot be definitively tested empirically until, with the passage of time, we reach the reference date. Even using Bayesian techniques to improve estimations with each new piece of data carries an underlying assumption that the iterative trials are asymptotically and deterministically approaching a solution, an assumption that might be confounded at any moment. In short, the contingency inherent in induction creates significant challenges when attempting to predict future outcomes in a non-deterministic universe.

For policymakers stuck in the Enlightenment paradigm this creates a double-edged problem. First, for predictions to be reliable they must assume a deterministic domain that unfolds through a continuous linear process of cause and effect, notwithstanding that the accumulating evidence around them is that both the natural world, and even more so, the social world, are not structured in this way. Second, the practical business of politics is determinedly short term. Although democratic and totalitarian leaders are driven by different worldviews, the primary pressures on them all are short term; democrats are inescapably focused on the next election which is rarely more than five years away, while totalitarians tend to be fixated on the current five or ten year plan. This is not to say that they are not aware of long term issues and that they don't devote time and energy to them, but it is to recognise that when under pressure, as is often the case, it is the long term that is sacrificed in preference to the short term.

It is in the nature of politics that politicians think their own ideology is the foundation of the most desirable future. Their greatest imperative is therefore to retain power in order to deliver that future. Within the prevailing rules of decency, all other issues are subordinate, and in particular, impacts on distant future generations, who are only tenuously part of their moral community and not at all part of their active political community, will always be subordinated to the claims of the here and now.

The absence of deterministic and predictable systems is everywhere a critical challenge to the Enlightenment project in its attempts to confront the problem of climate change. The future trajectory of climate change is critically dependent on the future trajectory of the values of a vast array of variables; to name just four – global population growth, global economic growth, the rate of low carbon technology innovation and the speed at which it displaces high carbon technologies. We can make assumptions about these variables, we can devise alternative models of the future with different values for them, but the best we can do is to produce a probability distribution for different possible futures. In later chapters I will examine the issue of uncertainty in more detail, but for the moment it is sufficient to recognise that while a probability distribution may be helpful in identifying the most likely outcomes, it does not follow that the least likely ones will not come to pass. If the

Enlightenment paradigm is threatened by anything, it is the recognition that the future is path dependent and therefore not deterministic. The greatest surprise would be a future without surprises. All effects having causes does not imply that the causes of all effects are knowable sufficiently to build robust predictive causal chains prospectively, or even in many cases, retrospectively.

Where exogenous agents may intervene, or where there is path-dependency driven by random bifurcations, or where a system manifests self-organisation that creates properties not present in its constituent parts, so-called emergent properties, it becomes difficult, and sometimes impossible, to establish causal chains that predictably link causes and effects. These are all features of complex adaptive systems that will be examined in more detail in the next chapter but in summarising the limitations of reductionism it is helpful to map out the boundaries of its utility. Snowden comments that in complex systems: 'Cause and effect cannot be separated because they are intimately intertwined' (2002, 25). The Enlightenment paradigm, built on reductionism and induction applied in a deterministic world devoid of discontinuities, is crumbling in the face of the accumulating anomalies. Something more is needed but first it is necessary to illustrate some of those anomalies as they have begun to emerge across the social sciences in recent decades.

Predict and provide

Inappropriate application of reductionism is indicated in situations where there is insufficient robust empirical evidence from which to construct any reasonable causal link between policy and outcome. This could arise for many reasons. For example, the situation might be novel, or although not novel, the policy is extended spatially or temporally into territory where there is little (or no) empirical evidence on which to establish the necessary causal links. It may also happen where vested interests influence decision makers to give greater weight to their views despite them not being socially optimal. In such cases, the policy will be determined by the opinions of the policymakers, however they may have been formed, and as a consequence are more a subjective reflection of their values than the outcome of an objective inductive application of reductionist principles (McCright and Dunlap 2010).

Inappropriate application of reductionism is also indicated where there are emergent properties whose appearance and effects cannot be predicted. As Funtowicz and Ravetz have observed:

> The problem is not simply that analysts are uncertain about the probability that some set of well characterized impacts will or will not occur. Instead, cumulative experience [...] has led to the understanding that the human perturbation of complex, nonlinear systems can induce outcomes that were *not even conceptualized before they actually occurred.*
>
> (Funtowicz and Ravetz 1990, cited by Howarth 2003, 261, emphasis added)

This is well illustrated by the UK's post-WWII 'predict and provide' transport strategy. This proposed that road traffic congestion could be avoided by predicting traffic growth and building roads to meet the anticipated demand. By the arrival of the new millennium this policy had been abandoned because '[t]he level of forecast infrastructure needed to meet an unconstrained growth assumption was seen as unsustainable both environmentally and financially' (Noland and Lem 2002, 18). The 'predict and provide' model failed to account for the new travel behaviours that were induced by the new roads, the emergent behaviours, that resulted in traffic growth always outstripping increases in road capacity. The failure of 'predict and provide' exemplifies a limitation in the applicability of reductionist thinking, its inability to cope with reflexive behaviours.[4] The tendency is always to underestimate the potential of positive feedback. Whether it is roads, health care, education, it is little surprise that providing easy and cheap access to public goods will increase demand.

Predict and provide is a policy orientation that is focused on resource provision, such as roads, schools, and public health services. By contrast, in determining the most appropriate policy responses to climate change, policymakers are more concerned with control mechanisms than service or infrastructure provision. Controls to limit emissions, controls to increase energy efficiency, and controls to encourage the development of low carbon substitute technologies are some of the mechanisms that policymakers are considering in their responses to climate change. In today's predominantly neo-classical global economy, governments are more likely to implement policies to provoke the market to deliver substantive action than to take direct responsibility for delivery themselves. I shall refer to 'predict and *control*' as opposed to 'predict and *provide*' to reflect this subtle but significantly different focus in climate change policy.

These climate change controls are located firmly within the problem-solving paradigm. They proceed from the simple, and superficially logical assumptions that anything that serves to reduce carbon emissions must be helpful and, assuming linearity, the more the reductions the greater the likelihood of success. But they also assume that reducing emissions does not have major negative consequences that also need to be taken into account (we shall shortly see that Beck would refer to these as 'reflexive risks'). It is not that policymakers are incapable of understanding that any policy might have some undesirable unintended consequences, but more significantly, that while some such consequences may be foreseeable, many will not be. This unavoidable ignorance, once unveiled by events, becomes a source of new policy responses that bring with them more reasons to intervene, continuously reproducing new undesirable unintended epiphenomena. This structure is far from the reductionist linear cause and effect in which one supposedly needs only to establish the causal chain in order to solve the problem.

If these controls are to be the sole policy tools, it is implicit that it is possible by such measures to reduce emissions on a scale and at a speed sufficient to avert dangerous climate change. For this assumption to be valid requires both a clear understanding of what constitutes 'dangerous climate change', and credible causal chains that link policy options to its avoidance. While policymakers are likely to be right that any reduction in emissions would be helpful, scale and non-linearity matter. For example, a common criticism of the Kyoto Protocol is that it is structurally incapable of producing significant emissions abatement, and, moreover, that this failure would come at a disproportionately high cost (Lomborg 2001; Prins and Rayner 2007; Tol 1999). In recent decades several concepts have emerged in the social sciences that identify various non-linearities in natural and social systems. Taken together they indicate a turn towards complexity. In the following sections I illustrate this turn by reference to four such concepts. The first is wickedness.

Wickedness

Urban planners Rittel and Webber (1973) used the terms *tame* and *wicked* to distinguish two different classes of problems. Wicked problems are public policy problems that defy solution. Tame problems are simply those that are not wicked. To be wicked, a problem must meet a number of criteria including most importantly that it has neither a definitive formulation nor a 'stopping rule' to know when it has been solved, if indeed a solution exists at all. Moreover, for wicked problems 'every trial counts' because:

> every implemented solution is consequential. It leaves "traces" that cannot be undone. One cannot build a freeway to see how it works, and then easily correct it after unsatisfactory performance. Large public-works are effectively irreversible, and the consequences they generate have long half-lives.
>
> (Rittel and Webber 1973, 163)

The notion of traces will become extremely significant when we later return to the practicalities of geoengineering. The traces mean that in an heuristic process, each trial starts from a different place because of the accumulated effects of earlier trials. Crucially, this distinguishes them from laboratory based experiments, small scale field trials and computer modelling techniques in which earlier trials, while perhaps providing useful data and insights that might inform subsequent efforts, do not alter the external context in which the trials are occurring. They can therefore be continually repeated as a means of corroborating results. For wicked problems each new trial is undertaken in a new reality; repetition and corroboration are not possible.

The current UK debate about the expansion of Heathrow airport is a perfect example of a wicked problem. Whatever option is chosen will alter social behaviours on a grand scale: local employment, house prices, ground transport infrastructure, public and commercial travel preferences and many other factors that have significant effects on people and the way they live their lives. Whatever decision is eventually made, the changes become embedded in a new reality such that when in due course it transpires that the chosen option is not optimific, as will inevitably be the case, it will not be possible to start again from the prior position. Moreover, that new reality will most probably have changed the nature of the problem, introducing new challenges that were previously either absent or not recognised, and rendering irrelevant others that were previously significant.

Rittel and Webber also argued that the principle of Popperian falsification cannot apply to wicked problems because they do not have solutions that are either right or wrong, but rather ones that are better or worse. This difficulty is compounded by the plurality of the publics likely to be affected and the wide range of possibly irreconcilable, yet reasonable, opinions about the degree of goodness or badness of the solution. As they put it:

> We have come to realize that the concept of *the* social product is not very meaningful; possibly there is no aggregate measure for the welfare of a highly diversified society, if this measure is claimed to be objective and non-partisan. Social science has simply been unable to uncover a social-welfare function that would suggest that decisions would contribute to a societally best state. Instead, we have had to rely upon the axioms of individualism that underlie economic and political theory, deducing, in effect, that the *larger-public* welfare derives from summation of individualistic choices. And yet, we know that *this* is not necessarily so.
> (Rittel and Webber 1973, 168, emphasis in original)

Rayner (2006) maps the notion of wickedness onto climate change showing that it fulfils all Rittel and Webber's conditions. He distils their ten criteria to six. First, a problem is wicked if it is difficult to formulate because it occurs on many levels. This is certainly true of inadvertent anthropogenic climate change with its two primary physical manifestations of global warming and ocean acidification that in turn have a broad spectrum of effects on every part of the ecosphere. Geoengineering is intentional climate change and therefore shares its meta-features.

Second, the problem of traces means that reductionist iteration is not available. While it is possible to conceive of very small-scale geoengineering implementations for research purposes, there are concerns about whether their effects would be measurable above the everyday noise of the ecosphere. If, in an attempt to overcome this noise, larger scale tests were contemplated, these could leave traces of the kind referred to by Rittel and Webber. Even if their enduring physical impacts on the global climate were insignificant, some argue that the long-term impacts on the human psyche of its changed relationship with nature would be potentially very significant. Tribe (1973), for example, eloquently questions the impacts on our psyche of a manufactured environment by reflections on the advent of plastic trees. Third, wicked problems are beset by what Rayner calls 'redistributive' implications due to irreconcilable conflicts between different interest groups about desired outcomes: the North/South split, generational splits, and so on. Different communities may have legitimate but conflicting needs. This same dilemma applies to all forms of geoengineering because their effects are temporally and spatially heterogeneous.

Fourth and closely allied to the issue of conflicting needs, Rayner identifies the problem of 'contradictory certitudes'. All social problems arise within a cultural context in which the community members have largely (although not entirely) shared traditions and beliefs about ethics and values. Rayner, displaying his anthropological stripes, asks the following questions to illustrate such certitudes: 'Are people fundamentally good? Are they evil? Is nature fragile and delicate? Is nature forgiving?' For each social group, these shared beliefs are certainties, core beliefs systems, their paradigmatic ideologies, that give the society its cohesion. Wicked problems engage with these ideologies, often provoking irreconcilable differences rooted in their diverse cultural heritages from across extended spatial and temporal scales. In these terms, climate change and responses to it, including geoengineering, are demonstrably wicked.

Fifth, wicked problems are not amenable to resolution; as Rittel and Webber put it, there is no stopping condition. Having applied the remedy, the problem doesn't go away, rather it morphs into or generates a new problem, much as contemplated in Beck's reflexive modernisation (to be considered in the next section). This also applies to geoengineering because for the reasons already discussed in terms of distributional implications and contradictory certitudes, there can be no universal agreement about either the set of parameters that defines the stopping point or their individual values. Finally, a key feature of wickedness is that the definition of the problem to some degree predetermines the nature of the solution – as noted earlier, framing matters.

Reflexive modernisation

The central thesis of Beck's seminal book *Risk Society* (1992) is that an emergent property of modernisation is that modern technologies create risks whose control demands the development of more technologies that introduce further risks, so creating a cascade of technologies whose purpose is to limit the negative effects of earlier technological progress. Moreover, the social context in which these risks are produced and managed is also altered by feedbacks, largely unintended, arising in reaction to that progress and the management of its co-produced risks, a process he terms *reflexive modernization*.

For Beck modernisation is broadly defined as:

> surges of technological rationalization and changes in work and organization, but beyond that it includes much more: the change in societal

> characteristics and normal biographies, changes of lifestyle and forms of love, change in structures of power and influence, in the forms of political repression and participation, in views of reality and in norms of knowledge.
>
> (Beck 1992, 50 n. 1)

Reflexivity in social systems refers to circular relationships between cause and effect that, after some iterations, not only render the causes and effects somewhat opaque but also begin to alter the social structures that previously prevailed. This reflexivity is apparent in many strands of social progress. Beck cites how the accumulating risks from globalised industrialisation are undermining the nation-state and the nuclear family that were instrumental in their establishment and initial prosperity; and how the creation of the welfare state, born out of a political ideal that favoured equality and abhorred poverty, has gradually begun to undermine the societies in which it has been most successful. This is well illustrated by a current debate in the UK in which some politicians and media represent the welfare state as a growing and increasingly unfair and unsustainable burden on the hard-working, and an abuse by the welfare-dependent.[5] They argue that this dependency has caused them to shun work in favour of state benefits that they opportunistically and immorally maximise by having ever more children thus reproducing a new poor, repeating the same cycle of welfare dependency learnt from their parents. Over time this process polarises society in exactly the opposite way to the intentions of the founding logic of the welfare state.

For Beck, reflexivity is a necessary feature of open-ended non-deterministic systems that manifest learning or adaptation. Actors can avail themselves of the opportunities offered by society but adapt their uses in ways unintended by the agencies providing those opportunities. The welfare state unwittingly encourages welfare dependency that in turn changes attitudes towards the welfare state. Advanced medicine uses science to combat ailments so encouraging lifestyles that endanger health – obesity and excessive use of antibiotics to name but two from a long and growing list. Beck also cites fertility science that he argues enables single parent families to become a matter of choice, so creating family units without a father and with unknown long-term social consequences if the practice were to become widespread. Although not one of Beck's examples, the Haber-Bosch process has enabled the widespread use of inorganic fertilisers that have vastly increased food production enabling global agriculture to feed a population that has almost tripled in 60 years. However, the run-off of excess fertiliser into rivers and underground aquifers has created a number of considerable environmental hazards including eutrophication, soil acidification, heavy metal accumulation and greenhouse gas emissions (Sulston 2012, 75). These responses to technological progress all entail threats to the established social order; they constitute the risks of Beck's *Risk Society*.

Postnormal science

Funtowicz and Ravetz developed the notion of postnormal science to describe situations where science is a central element of social policymaking, yet the facts are uncertain, values are in dispute, stakes are high and action is urgent (Ravetz 2011). In his analysis of the Climategate scandal following the hacking of emails at the University of East Anglia immediately prior to the 2009 Copenhagen COP15 summit of the UNFCCC, Ravetz discusses how climate change fits well with these core concepts of postnormal science. He explains:

> We can understand the root cause of Climategate as a case of scientists constrained to attempt to do normal science in a post-normal situation. But climate change had never been a really 'normal' science, because the policy implications were always present and strong, even overwhelming. Indeed, if we look at the definition of 'post-normal science', we see how well it fits: facts uncertain, values in dispute, stakes high, and decisions urgent.
>
> (Ravetz 2011)

The 'normal' of postnormal refers to Kuhn's concept of normal science whereby scientists extend marginal areas of knowledge (Kuhn 1962). Kuhn explains that:

> Normal science, the activity in which most scientists inevitably spend almost all their time, is predicated on the assumption that the scientific community knows what the world is like. [...] research is a strenuous and devoted attempt to force nature into the conceptual boxes supplied by professional education.
>
> (Kuhn 1962, 5)

Ravetz is highlighting the traditional problem-solving dynamic of normal science that depends on an analytical reduction of a problem into its constituent parts on the assumption that not only can the whole be explained purely in terms of its parts, but also, in terms of the prevailing broad understanding of how nature works, what Kuhn referred to as a paradigm. Where there is uncertainty about any of the parts, the uncertainty, it is assumed, can be statistically quantified in ways that enable a bounded solution that approximates to some rationally justifiable version of the truth. However, the essence of postnormal science is that where the facts are uncertain, values in dispute, stakes high and action urgent:

> *invoking truth as the goal of science is a distraction, or even a diversion from real tasks. A more relevant and robust guiding principle is quality, understood as a contextual property of scientific information.*
>
> (Funtowicz and Ravetz 2003, 2, emphasis in original)

As if affirming Funtowicz and Ravetz's ideas, the underlying principles of postnormal science also appear in Beck's notion of the 'decision paradox' that arises because major threats make action more urgent and also amplify the significance of knowledge gaps, making decisions 'more impossible' (Beck 2008, 117).

As several have commented, climate change sits firmly in the realm of the *postnormal* (e.g. Bellamy *et al.* 2012; Dessai and Hulme 2004; Sardar 2010), and as noted above in relation to wickedness, geoengineering shares the meta-features of climate change and can also therefore be categorised as *postnormal* science.

Critical theory

The final strand in this section of alternative approaches to science is Cox's critical theory. Cox was a political scientist. In *Social Forces, States and World Orders: Beyond International Relations Theory* (1981) he is concerned with international relations generally and not with science or more specifically with climate change, which at that time was only beginning to emerge onto the international political agenda. Nevertheless climate change is a global issue and is now well established as a matter of international concern. I suggest that the general insights offered by Cox are equally applicable to international climate negotiations in which the underlying science is a significant factor and is itself negotiated because, in Funtowicz and Ravetz's terms, the facts are uncertain, values are in dispute, stakes are high and action is urgent.

Cox introduces a distinction between *problem-solving theory* and *critical theory* that turns on the former operating within the confines of the current structure whereas the latter looks from the perspective of alternative structures. Cox provides a theoretical justification for the aphorism attributed to Einstein that 'We cannot solve our problems with the same thinking we used when we created them'. At least in terms of framing and defining the problem, Cox's problem-solving is similar to Kuhn's normal science because both are based on the idea of working within the existing paradigm to find solutions to well-defined problems. Critical theory is a more appropriate response to Rittel and Webber's wicked problems that defy clear formulation and are not amenable to solution. As Cox notes in comparing and contrasting problem-solving and critical theory:

> The strength of the one is the weakness of the other. Because it deals with a changing reality, critical theory must continually adjust its concepts to the changing object it seeks to understand and explain. These concepts and the accompanying methods of enquiry seem to lack the precision that can be achieved by problem-solving theory, which posits a fixed order as its point of reference.
>
> (Cox 1981, 129)

Cox's reference to a 'fixed order' chimes with the assumed immutability and knowability of the Laws of Nature assumed in the Enlightenment paradigm – the observable reality that exists independently of the observer and is there waiting to be discovered. This supposed precision of problem-solving theory, however, rests upon a false premise – the social and political order is not fixed but is in constant flux (Cox 1981, 129).

Summary of relational approaches

The common thread running through wickedness, reflexive modernisation, postnormal science and critical theory is the recognition that the context in which large-scale social problems reside is constantly in flux and is altered reflexively by interventions to address them, as well as by internal feedbacks. These processes introduce new risks some of which will not be appreciated until they manifest, and on occasions, as Beck explains:

> The preventive measures against catastrophic risks themselves trigger catastrophic risks, which may in the end be even greater than the catastrophes to be prevented.
>
> (Beck 2008, 119)

This also echoes Funtowicz and Ravetz's observation, already cited, that human interventions can produce outcomes that 'were not even conceptualised before they actually occurred'.

Wickedness is primarily concerned with identifying problems that are unlikely to yield to reductionist thinking where the goal is to make things better rather than make them right. Reflexive modernisation focuses more on the reflexive relationship between interventions and the context in which they are applied stressing the accretion of risks. Beck proposes, as a response, 'the unbinding of politics', by which he intends the widening of public participation in confronting these challenges. Postnormal science also proposes the engagement of an extended peer community similarly to widen the social base of those involved in developing responses in order to avoid the myopia and bias of the expert community. Finally, in order to effect the necessary social transformations, Cox also advocates a bottom-up/outside-in approach, emphasising the importance of an heterogeneous mix of inputs to policy-making where the *bottom* implies large numbers of those affected, *outside* implies diversity and novelty in perspective, and *transformation* suggests a stable but far-from-equilibrium state (Cox 1981, 135).

Reflexivity, heterogeneity, emergence, non-linearity, path-dependency are amongst the key features that have driven these authors to look beyond reductionism to confront the challenges of social policymaking. These alternative perspectives suggest that attempts to resolve the wicked problem of climate change using the problem-solving techniques of normal science incorrectly frame it as soluble because they wrongly assume that more knowledge implies less uncertainty. They fail to account adequately for accumulating risk production and other unpredictable emergent properties from the process of climate change. Nevertheless, as we will see in later chapters, normal science is the paradigm in which the great majority of those in the geoengineering community work and this locks them into 'mere pattern prediction' (Hayek 1975, 4) inhibiting the emergence of responses with the potential to transcend those established behaviours.

Conclusion

This chapter argues that the view from wickedness, reflexive modernisation, postnormal science and critical theory marks a shift towards a systemic view of the nature of the climate challenges we face. Rittel and Webber note that wicked problems 'defy efforts to delineate their boundaries', and those attempting to resolve such problems are 'caught up in the ambiguity of their actual causal webs'. Moreover, their proposed solutions are 'judged against an array of different and contradicting scales' deriving from 'the *growing pluralism* of the contemporary publics' (Rittel and Webber 1973, 167). These are all axiomatic features of the complex systems we shall explore in the next chapter.

Despite their powerful insights, these four concepts, wickedness, reflexive modernisation, postnormal science and critical theory, are more focused on understanding the world in which we find ourselves and the inadequacy of the tools we currently employ to address the challenges we face, than they are on providing alternatives, except in the broadest possible terms. Rittel and Webber end their paper by declaring that perhaps the most wicked problem of all is the wickedness of how best to confront wicked problems (Rittel and Webber 1973, 169). Beck proposes a differential politics 'to unbind politics' so engaging wider communities in the political process, but he makes no suggestions about what this new politics should be doing, deciding, or valuing; he stops with process (Beck 1992, 231–235). In his later writings (Beck 2003; Beck 2007; Beck 2012; Beck and Sznaider 2006) he emphasises the centrality of cosmopolitanism in addressing the social tensions brought about by reflexive modernisation but never explains in any comprehensive manner how this cosmopolitanism is to come about. Funtowicz and Ravetz also end with recommendations about process and attitude. They conclude that 'the appropriate style will no longer be rigid demonstration, but inclusive dialogue' (Funtowicz and Ravetz 2003, 7). Finally, Cox's critical theory is also about process rather than content. Process is necessary but, I contend, not sufficient.

These four concepts of social change demonstrate the limitations of a reductionist approach to wicked problems, they represent an incipient paradigm, not yet coherent but whose shape is beginning to emerge. Complexity theory provides some tools for policymakers that are useful in making the shift from the normal to the postnormal, from problem-solving to critical theory, and in distinguishing wickedness from tameness, thereby increasing the likelihood that we can avert the catastrophic consequences of the accumulating risks generated at each turn of the modernisation screw.

Notes

1 The IPCC (IPCC 2014, Glossary) distinguishes between prediction and projection as follows: 'Climate projections are distinguished from climate predictions by their dependence on the emission/concentration/radiative forcing scenario used, which is in turn based on assumptions concerning, for example, future socioeconomic and technological developments that may or may not be realized.' This is not a distinction that is routinely observed either in climate science or social science literature. I use these words interchangeably in common with most others. Etymologically *projection* implies something thrown forward and therefore references the prior state, whereas *prediction* implies only a prior declaration of a future state. *Forecast* is the Anglo-Saxon equivalent of *projection*, while *prophesy* is imbued with a sense of divine intervention.

2 Homoeopathic dilutions are usually so high that they pass beyond Avogadro's number and there is only the remotest likelihood of even a single molecule of the original substance left in the diluent.

3 The Large Hadron Collider (LHC) is the world's largest and most powerful particle accelerator – refer to the CERN website at http://home.web.cern.ch/topics/large-hadron-collider.

4 Notwithstanding the lessons learned from this empirical evidence, recent decisions by the UK government indicate that *predict and provide* is being rehabilitated. See for example debates in *the Guardian*, the BBC and *Daily Mail* and available online at, respectively: www.guardian.co.uk/environment/georgemonbiot/2011/oct/06/road-building-plans-tory-government;http://news.bbc.co.uk/1/hi/uk/3056636.stm; and www.dailymail.co.uk/news/article-2210738/Plans-200-new-roads-threaten-Britains-precious-countryside.html. All accessed 19 November 2012.

5 See for example *Daily Mail* articles *Curing the cancer of welfare dependency* published 6 June 2012, available online at www.dailymail.co.uk/debate/article-2174074/Curing-cancer-welfare-dependency.html; and *Stop rewarding laziness and promote work ethic: UK has world's worst idlers, say Tory MPs* published 17 August 2012; and in *the Guardian*, although a more sympathetic view is taken, the reality of welfare dependency is not disputed in *In a system with winners and losers, you can't have equality of opportunity* published 6 January 2012, available online at www.guardian.co.uk/commentisfree/2012/jan/06/deborah-orr-welfare-winners-losers; all accessed 18 November 2012.

References

Alexander, James. 2014. 'The Major Ideologies of Liberalism, Socialism and Conservatism'. *Political Studies*. doi: 10.1111/1467-9248.12136.

Beck, Ulrich. 1992. *Risk Society: Towards a New Modernity*. Sage Publications Ltd.

Beck, Ulrich. 2003. 'Toward a New Critical Theory with a Cosmopolitan Intent'. *Constellations* 10(4): 453–468. doi:10.1046/j.1351–0487.2003.00347.x.

Beck, Ulrich. 2007. 'Beyond Class and Nation: Reframing Social Inequalities in a Globalizing world1'. *The British Journal of Sociology* 58(4): 679–705. doi:10.1111/j.1468–4446.2007.00171.x.

Beck, Ulrich. 2008. *World at Risk*. First Edition. Polity Press.
Beck, Ulrich. 2012. *Twenty Observations on a World in Turmoil*. Polity Press.
Beck, Ulrich, and Natan Sznaider. 2006. 'Unpacking Cosmopolitanism for the Social Sciences: A Research Agenda'. *The British Journal of Sociology* 57(1): 1–23. doi:10.1111/j.1468–4446.2006.00091.x.
Bellamy, Rob, Jason Chilvers, Naomi E. Vaughan, and Timothy M. Lenton. 2012. 'A Review of Climate Geoengineering Appraisals'. *Wiley Interdisciplinary Reviews: Climate Change* 3(6): 597–615.
Brigandt, I., and A. Love. 2008. 'Reductionism in Biology'. *Stanford Encyclopedia of Philosophy*.
Buck, Holly Jean. 2012. 'Geoengineering: Re-making Climate for Profit or Humanitarian Intervention?'. *Development and Change* 43(1): 253–270.
Charlesworth, Mark, and Chukwumerije Okereke. 2010. 'Policy Responses to Rapid Climate Change: An Epistemological Critique of Dominant Approaches'. *Global Environmental Change* 20(1): 121–129. doi:10.1016/j.gloenvcha.2009.09.001.
Cox, Robert W. 1981. 'Social Forces, States and World Orders: Beyond International Relations Theory'. *Millennium – Journal of International Studies* 10(2): 126–155. doi:10.1177/03058298810100020501.
Dessai, S., and Mike Hulme. 2004. 'Does Climate Adaptation Policy Need Probabilities?'. *Climate Policy* 4(2): 107–128.
Durand, Rodolphe, and Eero Vaara. 2006. *A True Competitive Advantage?: Reflections on Different Epistemological Approaches to Strategy Research*. Chambre de Commerce et d'Industrie de Paris. www.hec.edu/heccontent/download/4765/130926/version/2/file/CR838.pdf.
Fischer, Frank. 2003. *Reframing Public Policy*. Oxford University Press. www.oxfordscholarship.com/view/10.1093/019924264X.001.0001/acprof-9780199242641.
Fukuyama, Francis. 1989. 'The End of History?'. *The National Interest* 16. National Affairs, Inc.
Funtowicz, Silvio, and Jerome R. Ravetz. 1990. *Uncertainty and Quality in Science for Policy*. Springer.
Funtowicz, Silvio, and Jerome R. Ravetz. 2003. 'Post-Normal Science'. *Online Encyclopedia of Ecological Economics*. International Society for Ecological Economics. www.Ecoeco.Org/publica/encyc.Htm.
Hayek, Freidrich. 1975. 'The Pretence of Knowledge'. *The Swedish Journal of Economics* 77(4): 433–442.
Howarth, R. B. 2003. 'Catastrophic Outcomes in the Economics of Climate Change'. *Climatic Change* 56(3): 257–263. doi:10.1023/A:1021735404333.
Kuhn, T. S. 1962. *The Structure of Scientific Revolutions 1*. University of Chicago Press.
Lomborg, Bjørn. 2001. 'The Truth about the Environment'. *The Economist* 4: 63–65.
McCright, Aaron M., and Riley E. Dunlap. 2010. 'Anti-Reflexivity The American Conservative Movement's Success in Undermining Climate Science and Policy'. *Theory, Culture & Society* 27(2–3): 100–133. doi:10.1177/0263276409356001.
Mörçol, Goktug. 2001. 'What Is Complexity Science? Postmodernist or Postpositivist?'. *Emergence* 3(1): 104–119.
Noland, R. B., and L. L. Lem. 2002. 'A Review of the Evidence for Induced Travel and Changes in Transportation and Environmental Policy in the US and the UK'. *Transportation Research Part D: Transport and Environment* 7(1): 1–26.
Prins, Gwyn, and Steve Rayner. 2007. 'The Wrong Trousers: Radically Rethinking Climate Policy'. John Martin Institute for Science and Civilisation. www.sbs.ox.

ac.uk/NR/rdonlyres/46EF88CF-64E0-4FA3-A0C7-AE5C3ED44E57/0/TheWrong Trousers.pdf.

Ravetz, Jerome R. 2011. '"Climategate" and the Maturing of Post-Normal Science'. *Futures* 43(2): 149–157.

Rayner, Steve. 2006. 'Wicked Problems, Clumsy Solutions: Diagnoses and Prescriptions for Environmental Ills'. Presented at the Jack Beale Memorial Lecture on Global Environment, University of New South Wales, Sydney, Australia.

Rittel, Horst W. J., and Melvin M. Webber. 1973. 'Dilemmas in a General Theory of Planning'. *Policy Sciences* 4(2): 155–169.

Sardar, Z. 2010. 'Welcome to Postnormal Times'. *Futures* 42(5): 435–444.

Snowden, D. 2002. 'Complex Acts of Knowing: Paradox and Descriptive Self-Awareness'. *Journal of Knowledge Management* 6(2): 100–111.

Sulston, Sir John. 2012. *People and the Planet*. The Royal Society Science Policy Centre.

Tol, Richard S. J. 1999. 'Kyoto, Efficiency, and Cost-Effectiveness: Applications of FUND'. *The Energy Journal*: 131–156, 397.

Tribe, L. H. 1973. 'Ways Not to Think About Plastic Trees: New Foundations for Environmental Law'. *Yale Law Journal* 83: 1315.

US Department of Defense. 2014. *Climate Change Adaptation Roadmap, 2014*. Available online from www.acq.osd.mil/ie/download/CCARprint_wForeword_c.pdf, accessed 19 February 2015.

4 Systems thinking

By comparison to reductionism, systems thinking is a recent innovation. Although its roots can be traced back at least to Aristotle to whom is attributed the dictum that 'the whole is greater than the sum of its parts', the modern father of systems thinking is von Bertalanffy, an Austrian biologist who first set out his principles of open systems in the 1920s (von Bertalanffy 1972). His central thesis was that from the early days of the Enlightenment, science had focused on closed physical systems but:

> [s]ince the fundamental character of the living thing is its organization, the customary investigation of the single parts and processes cannot provide a complete explanation of the vital phenomena.
>
> (von Bertalanffy 1972, p. 410, citing his earlier work from the 1920s)

His ideas were quickly taken up across a range of disciplines including cybernetics and economics. He observes them being applied to the social sciences in the 1960s and 1970s by Harvey, Laszlo and Demerath (von Bertalanffy 1972, 422). Fuenmayor (1991, 3) locates the further development of systems thinking in the works of three management scientists, Churchman, Ackoff, and Checkland from the late 1960s to the early 1980s. Since then it has been developed and extended as scientific methods have increasingly been applied to the understanding, development and management of a range of complex social systems in both public policy and private enterprise (Hoffmann and Riley, Jr 2002).

In the social sciences there are those that have examined systems thinking as a methodology for approaching social science studies in general; there are those that have written on climate change without referencing systems thinking but nevertheless embodying many systems concepts in other language; and there are a few who have begun reflecting on what systems thinking might have to offer in the formulation of public policy in general and geoengineering in particular. We will come to these in Chapter 6, but first it is necessary to examine complex adaptive systems and systems thinking in more detail.

From the now considerable literature on complex adaptive systems I focus on the work of Gunderson and Holling, not least because Holling is specifically referenced in the 1992 NAS report on climate change and geoengineering, one of the earliest authoritative papers on the subject (NAS 1992, 194 and 529), and their collaboration features in more recently published material (including some forty references in IPCC AR5). Their ideas on *panarchy* are regularly referred to as foundational concepts (e.g. Kinzig *et al.* 2006; Halsnæs *et al.* 2007, para. 2.1.4; Duit and Galaz 2008; Rockström *et al.* 2009; J. Walker and Cooper 2011).

Systems thinking

Systems thinking is the organisational principle embedded within the inchoate science of complexity. The IPCC defines systems thinking as 'a loose term' covering 'a growing body of concepts and models, which explores reality from different angles and in a variety of contexts'. They add that it has emerged precisely because of 'the inability of normal disciplinary science' to cope with the challenges of climate change and environmental sustainability (IPCC 2008, sec. 2.1.4). More briefly Capra defines it as 'thinking in terms of relationships, patterns, processes and context' (2005, 33) and von Bertalanffy as enabling the 'scientific exploration of wholes' (von Bertalanffy 1972, 415). In an attempt to answer the question '*What is complexity science*' Medd (2001) recognises the profusion of criteria that others have used to characterise systems thinking, in effect endorsing the IPCC's reference to it being 'a loose term'. However, he selects *emergence* as being central to all descriptions of complexity science. Emergence captures the notion that systems display properties that are not present in their constituent parts, or put another way, that they produce outcomes that are not deterministically derived and therefore can be neither deduced nor induced from an analysis of those parts. Emergence is a process by which some systems self-organise, thereby creating novel features; these are *complex adaptive systems* (CASs).

Emergence also applies to social systems. Walby focuses on emergent properties, contrasting the capacity that systems thinking has to account for them with the incapacity of reductionism to do so. She notes the potential that the concept of emergence has to address the relationship between different levels allowing much greater flexibility in understanding the interactions between individuals, and between groups of individuals, however these are constituted (Walby 2007, 462–463). This allows systems thinking to accommodate individuals of all types, including human and non-human, animate and inanimate.

From an overview of definitions of what constitutes a 'system', Ricigliano and Chigas conclude that there is general agreement that:

> a system consists of elements or parts, the links and interrelationships between the parts that hold them together, and a boundary, or the limit that defines what is inside or outside the system.
>
> (Ricigliano and Chigas 2011, 2)

Complex systems differ from *complicated* systems by virtue of their emergent properties. Complicated systems have their outcomes designed in and if properly constructed and operated, any given input to a complicated system will generate a predictable output (for example a pendulum, a car, a nuclear reactor). CASs, on the other hand, have emergent properties that over time produce evermore diverse and unpredictable outcomes. Darwinian evolution is a paradigmatic example of this process.

CASs comprise a large number of constituent entities displaying nonlinear and reflexive interactivity in which cause and effect can be difficult if not impossible to discern (Richardson, Cilliers and Lissack 2001, 7). Complex systems abound in most social structures and ecosystems at all scales. Figure 4.1 is a typical system map, in this case depicting the UK Government Office of Science's attempt to explain the obesity problem. Even a cursory examination of this diagram illustrates well the realities of reflexivity, indeterminacy and unpredictability. For any given input or combination of inputs in this system, no theory will support a well-defined causal chain enabling the outcome from those inputs to be predicted.

The increasing focus on complexity science in the academic environmental literature, beginning in the 1960s, has engendered much discussion about what constitutes a CAS. While Medd observes that complexity science is itself a CAS and therefore also in constant evolution, Levin (1998) distils earlier definitions to three core features: i) sustained diversity and individuality of components; ii) localised interactions among those components; and iii) an autonomous process, based on the results of local interactions, that selects emergent properties for replication or enhancement. These features combine to produce a dynamic system in which:

> the dispersed and local nature of an autonomous selection process assures *continual adaptation, the absence of a global controller*, and *the emergence of hierarchical organization.*

And crucially, he adds:

> The maintenance of diversity and individuality of components implies *the generation of perpetual novelty*, and *far-from-equilibrium dynamics.*
>
> (Levin 1998, 432, emphasis added)

When applying generalised CAS theory to the management of ecological CASs 'the key to resilience [...] is in the maintenance of heterogeneity, the essential variation that enables adaptation' (Levin 1998, 435). Although Levin's paper is concerned primarily with the impact that humans have on biodiversity, his theoretical perspective, and in particular the relationship between resilience and heterogeneity, are readily extended to encompass socio-ecological CASs more broadly.

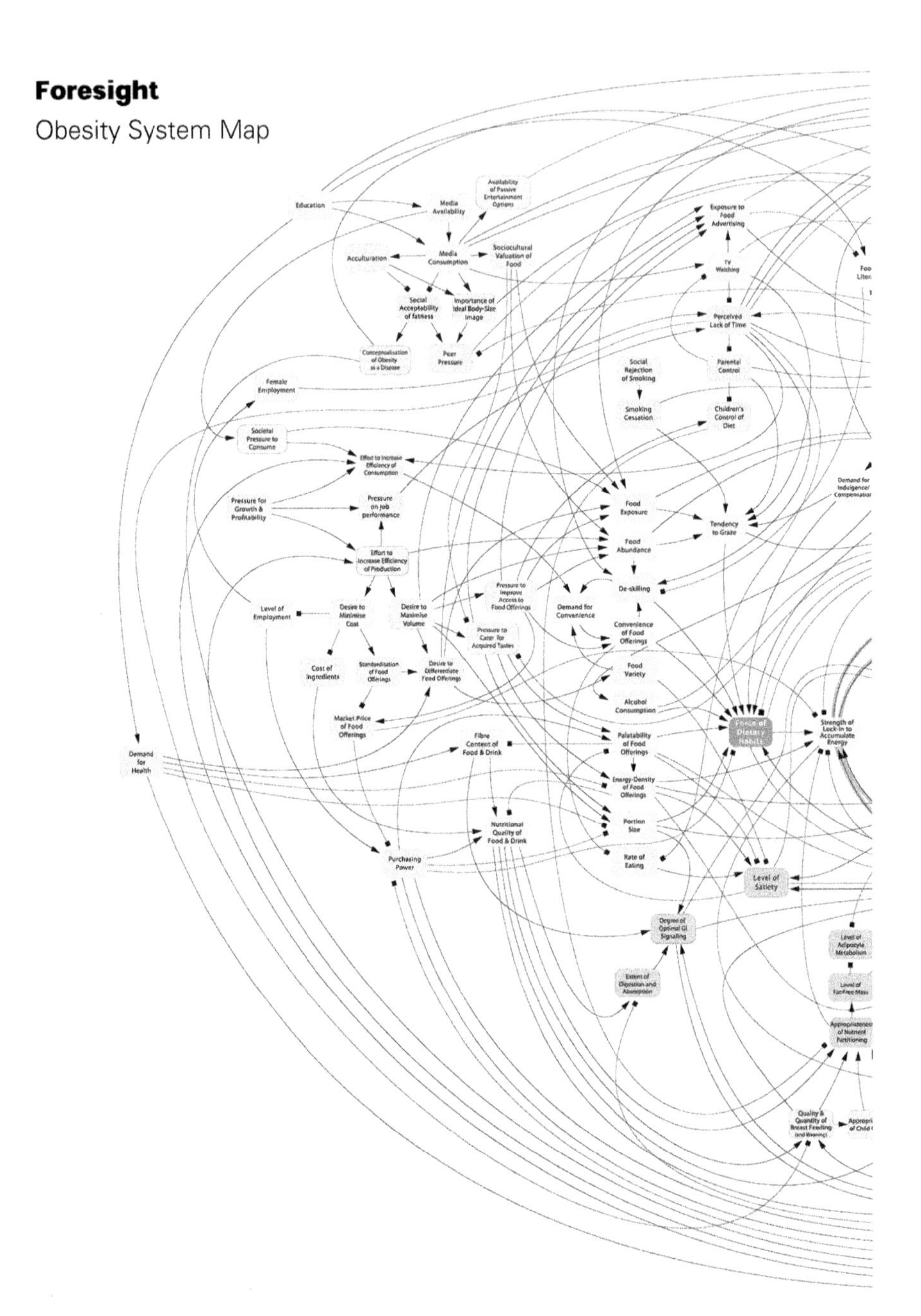

Figure 4.1 Obesity systems map
Source: prepared by Foresight and UK Government Office for Science.[1]

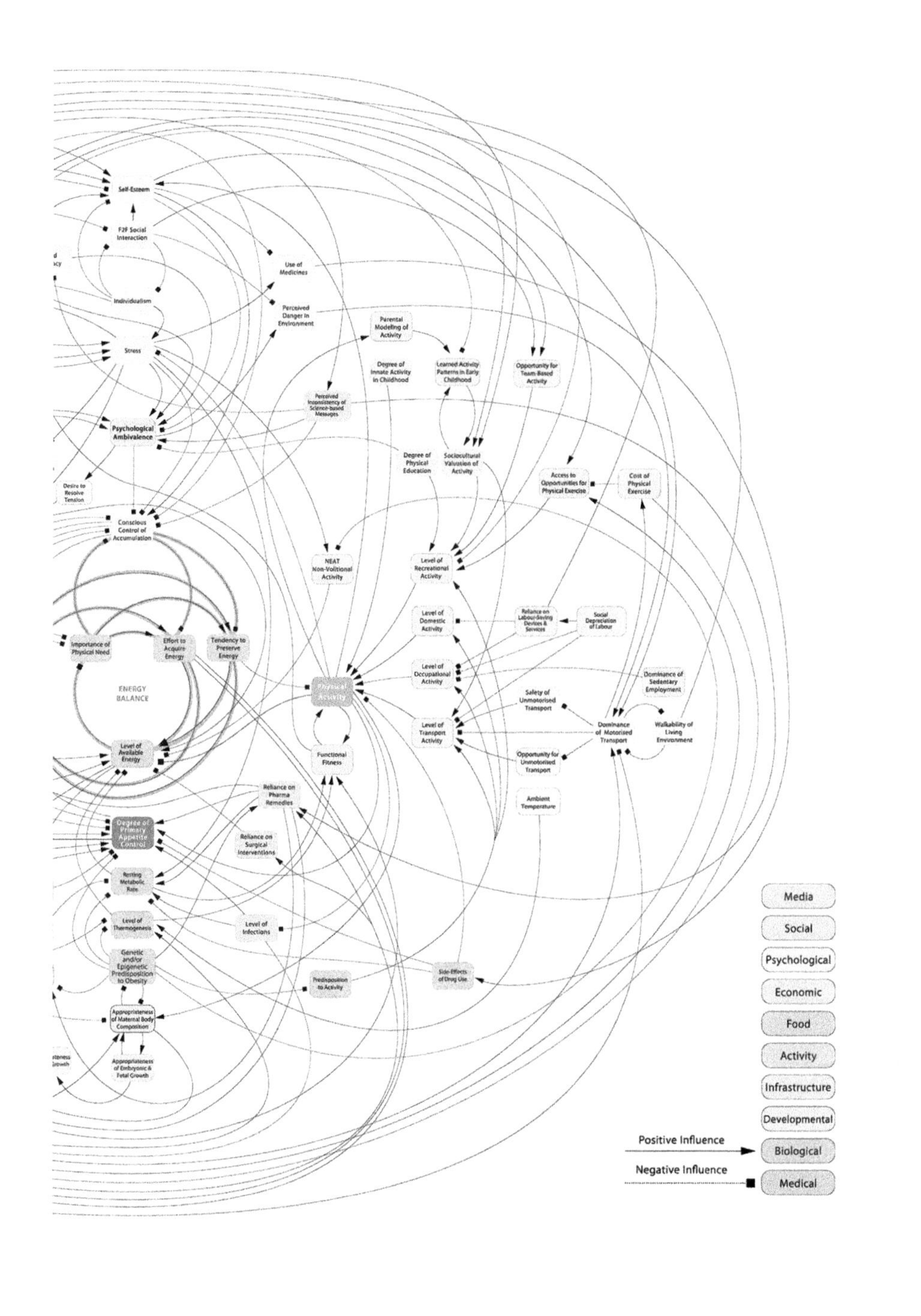

Self-Esteem
F2F Social Interaction
Use of Medicines
Individualism
Perceived Danger in Environment
Parental Modeling of Activity
Stress
Degree of Innate Activity in Childhood
Learned Activity Patterns in Early Childhood
Opportunity for Team-Based Activity
Psychological Ambivalence
Degree of Physical Education
Sociocultural Valuation of Activity
Desire to Resolve Tension
Access to Opportunities for Physical Exercise
Cost of Physical Exercise
Conscious Control of Accumulation
NEAT Non-Volitional Activity
Level of Recreational Activity
Level of Domestic Activity
Reliance on Labour-Saving Devices & Services
Social Depreciation of Labour
Importance of Physical Need
Effort to Acquire Energy
Tendency to Preserve Energy
ENERGY BALANCE
Physical Activity
Level of Occupational Activity
Dominance of Sedentary Employment
Safety of Unmotorised Transport
Level of Transport Activity
Dominance of Motorised Transport
Walkability of Living Environment
Level of Available Energy
Functional Fitness
Opportunity for Unmotorised Transport
Reliance on Pharma Remedies
Ambient Temperature
Degree of Primary Appetite Control
Reliance on Surgical Interventions
Resting Metabolic Rate
Level of Infections
Level of Thermogenesis
Genetic and/or Epigenetic Predisposition to Obesity
Side-Effects of Drug Use
Predisposition to Activity
Appropriateness of Maternal Body Composition
Appropriateness of Embryonic & Fetal Growth
Media
Social
Psychological
Economic
Food
Activity
Infrastructure
Developmental
Positive Influence
Biological
Negative Influence
Medical

From reductionism to systems thinking

Whereas reductionism requires the building of credible long-term causal chains to distant future local states, CAS theory illustrates the futility of this process and suggests other policy approaches to accommodate risk, uncertainty and unpredictability. Holling summarises this neatly when he exhorts us to:

> Embrace uncertainty and unpredictability. Surprise and structural change are inevitable in systems of people and nature.
>
> (Holling 2001, 391)

The parallels between Rittel and Webber's concept of wickedness and CAS theory are identified in their conclusion quoted below and highlighted by my annotations in square brackets:

> the problems that planners must deal with are wicked and incorrigible ones, for *they defy efforts to delineate their boundaries* [open non-deterministic systems] and to identify their causes, and thus to expose their problematic nature. The planner who works with open systems is caught up in *the ambiguity of their causal webs* [reflexivity and emergence]. Moreover, his would-be solutions are confounded by a still further set of dilemmas posed by the *growing pluralism* of the contemporary publics [many independent actors], whose valuations of his proposals are judged against an array of different and contradicting scales [heterogeneity].
>
> (Rittel and Webber 1973, 167, emphasis added and author's annotations in square brackets)

Similarly, I am not the first to notice the link between systems thinking and reflexive modernisation. For example, when reviewing theories of governance for problems that can only be dealt with over very long (multi-decadal) timescales, Loorbach notes that reflexive modernisation requires their adverse side-effects to be accounted for by allowing a continuous redefinition of their dynamics through a process that 'is characterised by diversity, uncertainty, heterogeneity of society, and the decreased possibilities for inducing long-term change by government' (Loorbach 2009, 166). These are precisely the dynamics that most challenge a conventional reductionist approach. By contrast, this continuous redefinition and reassessment are the essence of CAS theory and the domain of systems thinking.

Beck, although making no explicit reference to systems thinking, makes many references to policy approaches that demand 'variants that do not close off the future, but transform the modernization process itself into a *learning process*, in which the revisability of decisions makes possible the revocation of side effects discovered later' (Beck 1992, 178, emphasis in original). Indeed, for Beck the concept of reflexive modernisation emerged in response to his attempt to understand social processes that could not be analysed in conventional reductionist terms of linear cause and effect.

Beck's *learning process* echoes the heuristics at the heart of the autonomous process of renewal that drives emergence in complex adaptive systems. His *unbinding of politics* is a means whereby power is distributed more widely through society so that scientific and political elites no longer monopolise these decisions (Beck 1992, 192); it is tantamount to increasing heterogeneity and encouraging diverse multiple political interactions in an autonomous process of regeneration and renewal that allows for the heuristic selection of successful variants and the suppression of the unsuccessful – the essence of complex adaptivity.

Both systems thinking and reductionism employ iterative processes. However, there is a vital distinction between the two. Reductionist iteration proceeds by attempting continually to reduce the gap between actuality and some idealised optimal solution. It assumes a solution, an equilibrium position, and attempts to converge asymptotically ever closer on it. Conversely, the reflexive iteration of systems thinking assumes that there is no ideal or optimal solution but that systems display far-from equilibrium stability in which normality is constant change, mostly incremental but occasionally paroxysmic. Thus the iterative processes of systems thinking seek no more than to make continual incremental improvements while limiting the downside risks so that any unforeseen negative consequences are manageable. However, it doesn't deny the possibility of a system collapse that might defy our attempts at continual improvement. There is no teleology, just a hopefully rising saw-tooth of change in which, over time, the successes overwhelm the failures. Systems thinking acknowledges that some failures are inevitable and only by embracing them can we make progress towards a future whose characteristics cannot be defined too precisely in advance. Moreover, systems thinking recognises that some outcomes, at first understood as failures, may become critical to later successes; what constitutes 'failure' can also be normative and transient.

Panarchy

As already noted (Ricigliano and Chigas 2011), systems have boundaries. However, because everything in the universe is connected, the drawing of system boundaries becomes a matter of convenience to make it easier for us to understand them and how they interact with each other. Systems can be nested, with one entirely contained within another, they can be overlapping, or they can be entirely distinct, although in this case they will be indirectly connected by other systems because nothing can exist in isolation. In addition, entities can belong to more than one system, performing different functions in each. We all have multiple identities. A schoolteacher may also be a parent, a wife, a child, an Olympic athlete, an author, and so on, and each of these roles unfolds in different systems where she fulfils different functions. However, her participation in each of these systems impacts her participation in all the others. A central challenge of systems thinking is how to make sense of this unstructured infinitude of interconnected systems, distinguished from one another by indistinct, arbitrary and shifting boundaries. Gunderson and Holling addressed this question in their seminal book *Panarchy: Understanding transformations in human and natural systems* (2001) and other supporting journal articles (Levin 2003; Holling 1973; Holling 1978; Holling 2001; Levin 1998). To avoid what they considered to be the baggage associated with the word *hierarchy*, they created the neologism *panarchy*[2] to refer to these multiple interdependent adaptive cycles. Their theory turns on an understanding of the adaptive cycles that describe the dynamics of individual CASs, and the interaction between the adaptive cycles of multiple CASs.

From observations of many natural and human processes they discern a common four phase cycle that they describe as exploitation (r), conservation (K), release (Ω), and reorganisation (α) (Holling 2001).[3] They call these *adaptive cycles* and define them by three interacting properties: i) *potential* refers to the assets or capital inherent in the system that comprise the resources on which any change will draw and also limit its capacity to change; ii) *connectedness* describes the strength of the bonds between elements in the system and determines its internal controllability, it is a measure of the flexibility/rigidity within the system that is critical in shaping its response to perturbation; and iii) *resilience* is the system's adaptive capacity, and determines its ability to withstand unexpected or unpredictable shocks.

As the cycle moves from r to K it slowly accumulates capital and in so doing its elements become increasingly connected and ordered, in a process of creative construction. As the cycle peaks at K, the system becomes excessively connected to the point that it displays rigidity in the face of perturbation. This rigidity creates what Gunderson and Holling refer to as 'an accident waiting to happen' that in appropriately inauspicious circumstances will result in the bonds being forcibly broken, releasing the capital sequestered in the system as it rapidly collapses, passing from K to Ω and then into a new phase of renewal and reorganisation that begins at α. They use Schumpeter's term *creative destruction* to describe the transition from Ω to α. The ensuing process of renewal is also typically rapid as it transitions into another extended era of growth and consolidation as the cycle continues. In the regeneration some initiatives will fail and some capital may leak from the system or be acquired from other systems, as the establishment of a new order proceeds. It is also possible that the entire cycle is terminated in circumstances where the collapse is so total, and so much of the capital leaks away that reorganisation and renewal is no longer possible. Nevertheless, even in that situation, some of the released capital will be available to nourish other CASs.

The processes within CAS cycles must happen sequentially and the renewal and growth phases cannot both be optimised simultaneously. It may be that within a particular phase there are eddies at a smaller scale where nested sub-systems are flourishing or collapsing, so as to accelerate or retard the progress of the phase, but at the higher system level the meta-processes of renewal, growth, peaking and collapse will continue sequentially. Moreover, as the system passes through its cycle, at each stage it reflexively establishes the conditions for the next (Holling 2001, 395).

The third property, resilience, decreases as the connectedness increases, and so creates the vulnerability to an 'accident waiting to happen'. In the α-phase, when connectedness is low and resilience high there is less resistance to change, allowing the process of renewal and reorganisation to begin relatively unimpeded by forces of conservation. In the phase from Ω to α 'innovation occurs in pulses or surges, uncertainty is great, potential is high, and controls are weak'. This is a heuristic process, so while capital is accumulating during the reorganisation α-phase some arrangements will be sub-optimal. This may result in a temporary reduction in potential as the trajectory moves from initiation (α) to consolidation (r). Holling describes this as 'a time of both crisis and opportunity' (Holling 2001, 395).

A key feature of resilience is the creation of alternative pathways that allow system processes to continue after a major perturbation. These processes may, indeed probably will, undergo change as they shift from one path to another, but they do so continuously, without abrupt dislocation, so that the essential character of the original processes is preserved. Low to medium levels of connectedness can facilitate this process of adaptation in the face of shocks to the system, but high levels of connectedness imply rigidities that can undermine resilience by preventing processes from shifting from one path to another even where the escape route exists. The French Revolution illustrates this distinction well; it was well within the ability of the Ancien Régime to accommodate the demands of the common people in a way that would have averted a full scale revolution but the rigidities within their social order prevented them from doing so.

Gunderson and Holling note that this stylised adaptive cycle, while being typical of many CASs, is not seen universally, or at least it may be subject to significant variation in cases where a CAS remains in one or two phases much longer than the stylised version might suggest. They cite, for example, exploited arid rangelands and pelagic biotic communities that oscillate between the α and r-phases because they 'are strongly influenced by uncontrollable or unpredictable episodic external inputs and have little internal regulation' (Holling 2001, 396).

The rapidity of phase shifts around an adaptive cycle is tied to the system's spatial extension broadly in an inverse exponential relationship. CASs exist at many scales from the microscopic to the entire ecosphere itself and even beyond to the universe in its entirety following Clark's notion of the ex-*orbitant* (Clark 2010a). Consider, for example, a tree. A leaf emerges in the spring, flourishes through the summer, falls and decomposes in the autumn, and its organic components are partially reorganised into other living matter the following spring. This entire annual cycle takes place largely within the shade of the tree on which the leaf grows. However, the tree is also part of a landscape. That landscape will also change, perhaps dramatically due to seismic events, advancing and receding glaciations and climate change entailing desertification or new patterns of precipitation. Changes to the landscape, a much larger area than the shade of a tree, generally occur over geological time. However, even within large-scale long-term adaptive cycles, there are Ω-phases during which changes can be abrupt. There is, for example, strong evidence that some 14,500 years ago, sea level rose by 20m in less than 500 years (Deschamps *et al.* 2012).

From their appreciation of this spatial and temporal connection, Gunderson and Holling developed the idea of panarchy, a hierarchy of adaptive cycles with the smaller and faster ones at the bottom and bigger and slower ones above. The interaction between the CASs within a panarchy is such that the lower levels tend to destabilise those above while the higher ones tend to stabilise those below, in a mutually reinforcing process of reflexive adaptation (Levin 1998). This interaction is fuelled by the communication of a small quantity of information or material between adjacent levels. Holling describes this structure as 'a heuristic model, a fundamental unit that contributes to the understanding of the dynamics of complex systems from cells, to ecosystems, to societies, to cultures' (2001, 393). He illustrates these dynamics by reference to the way in which a forest stand moderates its local climate to narrow the range of temperatures experienced by its individual constituents. By extension it follows that, for human communities, individual cultural settings reflexively establish the norms that guide the behaviour and lifestyles of their members in a continuous process of adaptation wrought by both internal and external influences, a process in which diversity sustains the resilience of the system.

Changes influenced by exchanges between levels in the panarchy tend to occur in surges according to where each CAS is in its own adaptive cycle, with change being more rapid when the internal resilience of a system is low in its K to α-phases. Where multiple adaptive cycles at different levels have their K to Ω-phases in phase, these changes can be magnified even to the point that catastrophic changes can cascade rapidly upwards through one or more levels. For example, revolutions against totalitarian regimes consist of the mass of the population (lower/faster) rising up and destabilising an entrenched and oppressive elite (higher/slower) that is itself peaking and increasingly less resilient and therefore vulnerable to 'an accident waiting to happen'. It was just such a combination of circumstances that led to both the English and French Revolutions.

Conversely, a functioning democracy operates a safety valve allowing grassroots pressure for social change to be vented through periodic elections and a free press. This reduces the risk of a full scale revolution that would typically follow long periods during which voices at the lower level have been subjected to repressive forces from above. These voices, waiting for a propitious moment to rebel against the political elite, must wait a little longer before rigidities in the establishment (higher/slower) develop to a point where it becomes vulnerable to their uprising. These rigidities could manifest in inequalities in access to food, security, shelter, opportunity or any combination of these and other interests considered at the time to be of sufficient value that their denial is able to catalyse a shift from growth and accumulation to collapse. This panarchic structure explains why, even in democratic societies, practices that produce ever greater inequality cause increasing danger of structural instability, even during periods of relative plenty.

Uncertainty and unpredictability are central features of both the adaptive cycles of individual CASs and of their interactions with other cycles within the panarchy. While there may be increasing predictability as patterns become established during the long r-phase of growth and consolidation, the creative destruction of the release and reassortment phases, K through α to the beginnings of a new r-phase, is inherently unpredictable. Changes here are highly path dependent because at a time of low connectedness multiple bifurcations will arise where outcomes will be determined by volatile local conditions.

The central conclusion from this construction of complexity is that social policy that assumes that an ecosystem will remain permanently in the relatively stable r-phase, and in which pattern forming behaviours will improve predictability, is unlikely to provide for the shocks and creative novelty that emerge in the other phases or in interactions with other CASs. Even attempts to accommodate anticipated uncertainties such as through scenario planning depend upon the unfolding reality conforming to one of the scenarios planned. However, resilience to surprises cannot be pre-planned by surprise-specific resilience building policies because it is in the nature of surprises that they have not even been conceived of prior to their occurrence. Indeed, it could be argued that any event that has been planned for ceases to be a potential surprise, a subject discussed in more detail in the next chapter. It follows that only by a generalised attention to system resilience can the chances be improved for future generations to cope with whatever surprises befall them.

Understanding the ecosphere as a panarchy suggests that however resilience is to be bolstered in fulfilment of our obligations to distant future generations, it will largely happen in the growth and accumulation r-phase of the human social adaptive cycle and be undermined or threatened in the low resilience K to α-phases.[4] On this analysis, our long-term social policy goals, when seen through the lens of a panarchy of adaptive cycles, rely on extending the r-phase for as long as possible, because once the rigidities and vulnerabilities associated with the K-phase become dominant, there is increasing risk that the 'accident waiting to happen' will actually happen.

Panarchy and collective action

A panarchic framing can help understand the dynamics of many common collective action situations. As a negative example, consider Hardin's *Tragedy of the Commons* (1968) and as a positive one, the notion of symbiosis. In the *Tragedy of the Commons*, the actors, behaving entirely reasonably from their individual perspectives and in the absence of a higher authority (global controller), collectively destroy a common resource on which they all depend. There are many examples of this in practice including depleted fisheries and the virtual extinction of bison and blue whales (Hornaday 1889; Schneider and Pearce 2004). A panarchic understanding of these hunting and commercial whaling practices is that their respective instrumental value, once recognised, was developed and exploited in a process of growth and accumulation (r-phase) but as the respective stocks began to be depleted, the established practices were too deeply entrenched to yield to the meagre forces calling for conservation and as a result, they propelled themselves out of growth and accumulation (r), towards a peak (K) and collapse (Ω), releasing their accumulated assets into a faltering process of renewal (α).

The example of whaling illustrates particularly well the linear thinking of the forces of regulation and its perverse consequences. The League of Nations brought into effect in 1935 a Convention for the Regulation of Whaling that secured protection for Right and Gray whales and provided a mechanism for the collation of statistical information (Schneider and Pearce 2004). Initially they restricted the worldwide catch to 16,000 BWUs.[5] This had disastrous conservation effects as it encouraged the whalers to take only the largest individuals resulting in rapid depletion of stocks of the target species. In 1937 further restrictions were introduced to protect females accompanied by calves, and the closure of the southern waters north of 40°S. However, this had the effect of increasing the activity in the unrestricted areas. Immediately after WWII, the International Whaling Commission (IWC) was established but it had little impact in halting the decline of the targeted species. Quotas were set more in line with the whalers' short-term commercial needs than with a view to sustainability. There were also indications of significant under-reporting of whale catches. Not until 1972, responding to increasing political pressure, did the IWC drop the BWU in favour of quotas by species but by this time all the large whale species had been hunted almost to extinction and the catches were entirely made up of Minke, the smallest whales species. Finally due to many non-whaling nations joining the IWC the balance was shifted in favour of the anti-whalers and a moratorium was introduced in 1982 with effect from 1985/86. However, by this time the whaling industry had largely collapsed because the markets for most whale products had been replaced by synthetic alternatives. If the regulation of carbon emissions were to follow a similar trajectory as that for whaling, we will have to wait until 2030 before emissions are properly regulated, but by then the regulations will be redundant because the market will have largely replaced fossil fuels. However, by then the accumulated and latent damage from global warming will be severe and irreversible.

The collapse of the bison and whale populations terminated the industrial activities that supported them and the capital and human resources previously devoted to them were redistributed into other activities, some no doubt still involved in hunting or pelagic fishing as the assets (finance, men, guns, ships, and their accumulated skills) were redirected into new activities to contribute to a new Ω-phase of the adaptive cycle, while others leaked away into totally unrelated activities in parallel adaptive cycles.

In contrast to the Tragedy of the Commons, symbiosis is an ecological phenomenon that arises where two species flourish in parallel. These relationships can be commensal, where one benefits and the other is unaffected, or mutualistic, in which both benefit. There are many examples of such arrangements; for example whales and barnacles, egrets and cattle, ungulates and their intestinal bacteria. Symbiosis has been extended to describe social phenomena where two communities flourish in parallel, but whether commensal or mutualistic, they both retain their separate identities rather than merging or integrating. In a fascinating paper from 1928, Park discusses the *marginal man*, an individual who, as the result of migration, must grapple with the tensions of trying to live simultaneously in two diverse cultural groups, that from which he comes and that of his new hosts. He suggests that it is 'in the mind of the marginal man that the process of civilization is visibly going on, and [...] may best be studied' (Park 1928). He explains the commensal relationship between gypsies and hobos and the communities where they temporarily reside, and the mutualistic relationship between the Jewish ghettos of pre-Nazi Europe and their host communities.

From a panarchic perspective, while host and migrant communities are both in their growth and accumulation r-phases, the symbiosis works well. But, once the resilience of either community in relation to the presence of the other is tested, as one or both approaches a peak, the opportunity for conflict grows. One way in which this danger may be mitigated is for the marginal man to become interfused rather than merely intermixed with his host. However, in this event his independent identity becomes increasingly compromised. Where these symbiotic processes collapse, genocide often results: the Jews and the Nazis in Central Europe, the Hutu and Tutsi in Rwanda are just two recent examples from a depressingly long list.[6] Genocides may be understood as singularities, abrupt and painful changes that arise as a consequence of a failure of resilience at a systemic level. However, as the adaptive cycles continue, the disruption results in assets and potential being released and realigned in new combinations as people remake their lives. Some of these initiatives are successful, others not, some see the marginal man being integrated into the new order, others see him migrate to seek well-being elsewhere. But after a period of readjustment, a new stability arises and a new r-phase of growth and accumulation commences.

Perhaps the challenge for human civilisation, when expressed in terms of panarchy, is continuously to refresh the community's connectedness and resilience so as to remain indefinitely in the stable but far-from-equilibrium r-phase of growth and accumulation. Whereas a reductionist approach would construct this as an analytical problem-solving challenge, systems thinking recognises that the tensions between potential, connectedness and resilience create an ineffable present that has already moved to a new state before we have fully understood it. This defeats an observer/observed analysis and requires an alternative approach to establishing the best way forward. It also recognises that the best way forward may just be less bad than the alternatives. As will be explored in more detail in later chapters, systems thinking introduces some considerable methodological and epistemological challenges. However, the benefits offered by CAS theory have considerable potential to reframe the ever-increasing complexities that confront us in living sustainably and in harmony with each other and with planetary systems, such that the practical difficulties it introduces should not be allowed to undermine our commitment to it.

Systems theory in the social sciences

Even within the realms of CAS theory, there are different emphases according to the authors' worldviews. Whereas Prigogine focuses on the internal functioning of systems, Waldrop is more concerned with their external relations (Walby 2007, 456). Walby also comments (2007, 457) on the diversity of applications and theoretical contexts in which complexity theory is deployed by social scientists. Byrne is concerned with realism, Cilliers with postmodernism, Luhmann with phenomenology and Jessop with Marxism. Urry's focus is on mobilities and the global, De Landa's on non-linearity and Wynne positions 'complexity theory as a challenge to both reductionism and the denial of uncertainty among science policy makers'. Luhmann's failure to address 'power and inequality' and the empirical realities of geopolitics, drew considerable criticism (Beck and Sznaider 2006, 16; Walby 2007, 457) illustrating that whatever merits CAS theory may have, as a device for representing our world it is no less immune to critical scrutiny than any other system of thought. But notwithstanding this scrutiny, or indeed, perhaps because of it, CAS theory has increasingly penetrated the social sciences.

Reductionist accounts of the world are challenged by the occurrence of singularities, the abrupt irreversible changes that emerge so unpredictably in both natural and social domains. Reductionism struggles in a world where effects are not 'gradual and proportionate', pointing to the need for a theory that accounts for '[t]he smallness and perhaps contingency of the event that precipitates these largescale changes'. CAS theory provides a framework in which singularities are understood as natural processes and while their onset and outcomes might be unpredictable, they are nevertheless inevitable (Walby 2007, 464).

Complex adaptive systems theory is not the only systems theory. Walby discusses the challenges for social theory in accommodating multiple intersecting social inequalities, inequalities that co-exist and co-produce the environment they inhabit. She notes that negative connotations associated with the concept of *system* have provoked some social scientists to adopt other terms to describe this concept, such as *social relations*, *network*, *regime* and *discourse* (2007, 455). She also explores the development of systems theory and identifies several shortcomings of pre-complexity theory versions to 'the theorization of multiple sets of social relations in the same institutional domain' (2007, 454). Pre-complexity notions of systems assumed that their natural state was one of equilibrium to which they would return after small perturbations. They also assumed that systems were discrete entities that could be understood in isolation. The absence of complexity restricts their capacity to allow for: i) the system not to saturate its territory; ii) multiple sets of social relations to coexist without being within a nested hierarchy; iii) categories to be ontologically constituted in multiple domains; and iv) systems to not return naturally to equilibrium after being subject to small perturbations, and as a consequence failing to recognise the possibility of singularities. These limitations prevent pre-complexity theorisations of systems from grasping the incoherence and messiness of the real world.

For example, a government's writ may not necessarily prevail in all its lands, as political leaders throughout history could attest. In a globalised world, social relations increasingly arise both within and between geographically and socially diverse entities. International non-governmental organisations and transnational corporations are prime examples of entities that routinely establish relations that cross-cut and exist outside the structure of established national hierarchies. The density and diversity of relations between entities throughout the ecosphere inevitably results in similar emergents arising in multiple locations, if not simultaneously, at least in close temporal proximity. This analysis rests on a sense that events do not just arise in a chronological sequence, creating a linear history of 'one thing after another', but rather that they arise kairotically, a temporal notion that regards events as being of their time, of arising from propitious complex interactions in entirely unpredictable and non-deterministic ways (Hedaa and Törnroos 2001; Hedaa and Törnroos 2008; Miller 1992; Miller 1994). They acquire the status of memes by resonating with prevailing ideas, a resonance that allows them to attract and adhere in the production of emergent properties, which, when they catch the tide of social change, they replicate in multiple parallel contexts. Recent examples are the Arab Spring of civil unrest across the Middle East occurring more or less simultaneously in multiple separate locations; and even the 2008 financial crisis that would not have happened but for a globalised neo-liberal ethos that exalts individual freedom even at the expense of social cohesion. More positively, consider the widespread (but sadly not global) strides made in gay and ethnic minority rights during the last half century.

Finally, there are any number of events, of modest significance in themselves, that have catalysed extraordinary and irreversible social upheavals. Again, Mohamed Bouazizi's self-immolation in a personal protest against police corruption and ill-treatment in Tunisia that sparked the Arab Spring in 2010, and almost a century earlier, the assassination in Sarajevo of Archduke Ferdinand, illustrate forcefully how on occasions, consequences can extend vastly beyond the local significance of the events that precipitate them.[7] These examples epitomise how theorising social systems as complex adaptive systems allows them to intersect, yet have different spatial and temporal reaches, a conceptualisation that is essential in order to account for their co-evolution and reflexive path dependent co-production.

Apart from Walby's examination of complex adaptive systems theory as a framing for enquiries into the social sciences, other social scientists introduce systems thinking, if not unwittingly, then certainly without using the systems thinking lexicon. A special edition of *Theory, Culture & Society* (TCS) (Shove 2010), devoted to climate change, contains contributions from several luminaries of the world of human geography. Concepts of reflexivity and emergence are explicit in the writings of Shove, Beck, Hulme and McCright referring to the iterative interactions between system elements and their (unpredictable) emergent properties that become part of a continuously renewed reality, providing new knowledge, changing the circumstances and ultimately, moulding new normativities. Szerszynski offers the insight that rather than the technology being a response to the 'disinterestedness of science', climate science is in fact 'shaped' by the potential of the technology such that the climate science simultaneously alerts us to the threats of climate change and conditions our response to them. This is an implicit recognition of systemic framing expressed in social science, rather than in complexity theory terms. Embedded in this observation is the reflexivity in the relationship between actors making them at once both subject and object. Indeed, its focus on the *relationships between* elements within the system rather than on the elements themselves, is a significant component of a systems thinking approach.

The impact that reflexivity has on confounding chains of causality is highlighted by Clark, Hulme, Szerszynski and Wynne. Wynne for example, cautions that climate modelling 'should be received less as predictive truth-machine and more as reality based social and policy heuristic' (2010), and Hulme argues that the multiplicity of actors is the source of 'hybridity' in the weather and that 'the weather cannot be so forensically dissected into [...] causal elements' (2010). Szerszynski claims that 'climate science's action-orienting power' constrains an understanding of the metabolic relation of humanity and nature to narrow causal terms, presenting it 'as a problem to be solved rather than an opening to be responded to' (2010). He argues that if we are to engage with our environmental responsibilities, there must be an 'opening of the climate', releasing it 'from its technological incarceration'. Clark (2010b) focuses on the reflexivity of the heuristic response to the confounding of causality, referring to a reliance on 'experimentation and improvisation [as much as on] precaution and self-restraint'.

Similarly, the systems concept of heterogeneity appears throughout the TCS special edition although expressed in terms such as plurality, diversity, multiplicity, cosmopolitanism, 'binding humanity together in new relations of interdependence', and ambiguity. Emergence also travels through this literature, often under its own name, often not. For example, Shove refers to:

> transitions toward sustainability [that] depend on societal innovation in which contemporary rules of the game are eroded, in which the status quo is called into question and in which less resource-intensive regimes, routines, forms of know-how, conventions, markets and expectations take root.
>
> (Shove 2010)

It seems that CAS theory is being taken up almost osmotically within the social sciences, illustrating its power to account for those aspects of humanity's being that defeat reductionist approaches.

From systemic resilience to policy robustness

The central argument thus far has been that, in the quest for timely and effective policy responses to climate change a) the conventional scientific method makes undeliverable demands on predictive methodologies; and b) systems thinking offers complementary methods with the potential to overcome many of these limitations. The core challenge is to escape dependence on long-term prediction because:

> [a]ny policy carefully optimized to address a "best guess" forecast or well-understood risks may fail in the face of inevitable surprise.
>
> (Lempert, Popper and Bankes 2003, xiii)

In systems terms, the best protection against surprise is offered by resilience, the ability both to recover from shocks and to adapt to a changed environment. How do we identify those elements that impact resilience in order to better manage them? If we attend to some but ignore others, whether by choice or ignorance, how are we to know the significance of those others? In most complex systems there is redundancy that enhances resilience by providing alternative pathways in the event of a failure elsewhere in the system. The value of these redundant components, and even their identity, may not emerge until they are called upon. How do we strike the correct balance between efficiency and redundancy so as not to increase vulnerability by sacrificing apparently inefficient redundant components that, unbeknown to us, may prove crucial in the system's capacity to avert a future catastrophic collapse?

Policy robustness is the practical policy objective designed to confront this problem (Lempert, Popper, and Bankes 2003; Walker, Haasnoot, and Kwakkel 2013). Robustness here means specifically that:

> rather than seeking strategies that are optimal for some set of expectations about the long-term future, [policy robustness] seeks near-term strategies [...] that perform reasonably well compared to the alternatives across a wide range of plausible scenarios evaluated using the many

> value systems held by different parties to the decision. In practice, robust strategies are often adaptive; that is, they evolve over time in response to new information.
>
> (Lempert, Popper, and Bankes 2003, xiv)

This extract highlights four conditions for robustness, each of which is necessary but none individually sufficient. First, the many scenarios against which a policy option is assessed need be merely *plausible*, their probability and the credibility of the causal chains that might explain them are of little concern – we know with virtual certainty that highly improbable events will occur, we just have no way of knowing which ones. Second, the predicted policy outcomes need to be evaluated 'using the many value systems held by different parties' affected by the decision. This is necessary not only as a matter of political legitimacy but also because it enshrines within the policy process the heterogeneity essential to nourish the emergence of innovations that serve the system as a whole, rather than only addressing the needs of those with a voice.

Third, the criterion for acceptability is that the policy need perform only '*reasonably well* compared to the alternatives' and not that it be optimal for any subset of those alternatives. Robust strategies are strategies that are least likely to result in catastrophic failure across the widest range of plausible futures. Robustness abandons the goal of optimality and promotes survival. It assumes that if we protect against the downside, irrepressible human ingenuity will look after the upside. Moreover, it recognises that beyond our vital needs of food, shelter and security, what constitutes well-being is subject to continuous change making it increasingly challenging for any current generation to know what might be optimal for increasingly distant future generations.

Long-term policy optimality is a mirage. Assuming that the fundamental integrity of a complex adaptive system is not threatened, future members will cope with whatever confronts them and that will constitute their normality. The possibility that their predecessors might have acted differently to create a different and supposedly more desirable future from that which actually unfolds will be confined to the realms of 'if only' conjecturing. They will get on with their lives and like all generations before and after them, make the most of what they inherit and be more or less content with their lot. That the world might have been more or less populous, that people might have been more or less materially prosperous, that technology might have been more or less advanced will count for very little provided that the great majority has food, shelter, security and sufficient human dignity within whatever social structures then prevail. Long-term optimality as a policy objective only makes sense in an unchanging world. That is not the world we have.

Fourthly, robustness implies continuing openness to policy adaptation as conditions change, a concept Walker *et al.* refer to as *dynamic* robustness (2013). They distinguish it from *static* robustness in which no structural provision is made for the continuing review and revision of the policy. Adaptive policymaking requires a process to be in place *from the outset* to enable the continuous monitoring and reflection on progress that makes the policy adaptation routine rather than occurring sporadically on an *ad hoc* basis.

Another central feature of the robustness approach is that its purpose is to identify near-term strategies that serve long-term objectives. While recognising the intractable uncertainty associated with predicting the distant future, a robustness approach also recognises the need for timely action. It extols the value of embracing that uncertainty 'rather than spending large amounts of time and effort on trying to reduce it, and waiting to take action until the uncertainties have been resolved' (Walker, Haasnoot, and Kwakkel 2013, 971).

In summary, the core propositions that translate systems theory into policy practice are first that by pursuing policy robustness we maximise our chances of building resilience, and second that by building resilience we minimise the likelihood that systemic catastrophe might prevent future generations from flourishing according to their own lights.

Climate change and systems thinking

Armed with these insights, how might they be applied to geoengineering? The grand objective of climate change policy, as declared by UNFCCC, is 'to avert dangerous anthropogenic interference with the climate system'. We can reinterpret this objective in terms of panarchy to mean adopting policies that will prevent, or at least defer, the onset of the Ω-phase in the critical CASs upon which human flourishing depends. We know from Gunderson and Holling that this is achieved by maintaining their resilience. If a world without surprises is unimaginable, perhaps we can imagine a world in which, by building systemic resilience, surprises can be absorbed without courting a generalised catastrophe. This resilience is not so much the ability to recover quickly from some local disaster, important though that may be, but rather the continuous and widespread reproduction of that capacity by a zealous and generalised attention to the heterogeneity on which resilience depends. Climate change ceases to be a problem to be solved and is reframed as a situation to be continuously managed (Loveridge 2008). The best we can do is limited to the best we can do – no generation can be judged to have failed in its obligations to future generations by virtue of not having done that which was not within its power to do. Since all policies are inescapably sub-optimal, the pursuit of optimality tempts us to reach beyond our grasp. Rather, we need continuously to accommodate the shortfall between our achievements and our aspirations, by treating it not as failure, but as a routine and even positive aspect of a heuristic learning process.

Complex adaptive systems theory is already beginning to appear in the climate change debate. Adaptation, broadly defined, is a systems focused activity that is concerned with maintaining the system in its r-phase of growth and accumulation, and lowering its vulnerability to the collapse of the Ω-phase. Pelling and High (2005) survey a range of approaches to adaptation and identify the distinction between adaptations that 'reinforce existing organisational or system stability and those that modify institutions to add resilience through flexibility'. Evans makes a similar distinction between *ecological resilience*, that focuses on systemic change, and *social resilience* that 'tends to be more parochially concerned with recovery from disaster to an identical pre-existing state' (Evans 2011, n. 1). For these and many other authors, the increasing focus on resilience is a clear manifestation of the emergence of systems concepts in the climate change discourse, even when systems thinking is not explicitly acknowledged.

Pelling and High use social capital as a lens through which to study 'the production of adaptive capacity among collectives [at all scales]'. They list a number of attributes of 'resilient adaptation' that include distributed decision-making, diversification of inputs, learning from past experience, active experimentation and support for innovation (Pelling and High 2005, 309). These attributes readily map onto those of complex adaptive systems theory although Pelling and High do not make this link.

The shift from 'solving' to 'managing' emphasises that the future will not be settled and unchanging. Evans offers 'an account of the conceptual basis of adaptation in resilience ecology' (2011, 233). He approaches 'resilience and urban adaptation' through complex adaptive systems theory based on work from Gunderson, Holling, Folke and others. He stresses the capacity that systemic thinking has to overcome the limitations of 'traditional conceptions of scientific knowledge that were predicated upon knowing a world that was not a moving target'. In emphasising the different mindset implied in this shift away from 'traditional science' to resilience thinking, a shift that echoes Funtowicz and Ravetz's move from normal to postnormal science, Evans comments that:

> Modernist tropes of planning, managing and regulating the environment are thus replaced with governance-style metaphors of 'navigating', 'surfing' and even 'dancing' crises. [...] The central lesson of resilience thinking is that environmental managers should avoid optimising a system to one specific set of stable environmental conditions, as they will reduce the ability of the system to adapt when those conditions change.
>
> (Evans 2011, 229–230)

It is clear that the shift from reductionism to systems thinking[8] in addressing climate change has taken root within the social sciences in relation to adaptation. A key question is how to extend this from adaptation to geoengineering. Some tentative steps in this direction have been taken by Galaz (2012) and Hall and Pidgeon (2010) but these amount to little more than a recognition of the challenge such a step entails and a call to action. Turning these ideas into concrete policy formulations in which geoengineering is framed in terms of resilience and systems thinking, has yet to begin. Perhaps a place to start once we have recognised the wickedness of climate change and the post-normality of climate science, is to expect no more from geoengineering than that it be a palliative, and whether it is to be a prophylactic, emergency or terminal palliative is an open question, but it cannot be a cure.

Conclusion

Conceptualising climate change in panarchic terms opens up avenues of enquiry and understanding that might previously have remained obscure. Scaling down from the multiverse to the sub-atomic events that combine to determine our earthly lives, there is an infinity of systems and subsystems, some in nested hierarchies, others in parallel strands, but all connected and interacting with at least one other to produce the cosmic whole. These systems will have the hierarchical spatial and temporal characteristics described by Levin, with those operating over larger territories generally doing so exponentially more slowly than those whose physical domain is more confined. Despite the capacity for entities to belong to multiple CASs, each system will be largely, but crucially not entirely, independent of others. Indeed this quasi-independence is a key determinant of the constructed boundaries that separate it from other systems. Changing views about these relations may cause the boundaries between systems to be redefined, or for new systems to be conceptualised. The boundaries are often fuzzy because systems are intellectual constructs, not real entities, and we can redefine them to suit our purposes.

Interactions between systems often occur when systems are at different stages in their adaptive cycle. If they are in their stable r-phase, those higher in the panarchy will tend to stabilise those below that might be suffering from low resilience as they head towards their Ω-phase. However, if those above are themselves experiencing low resilience, provocations from below are likely to result in a tipping point that will overturn a long-established equilibrium and usher in the radically new. Whether the events that follow a tipping point are catastrophic or not is a normative question.

If CAS theory tells us to focus on resilience, we now have a direction of travel with the ideas discussed in this chapter as guides to light our way. Nevertheless there remain important unanswered questions. Which of the infinity of CASs above, below and alongside human society in the panarchic CAS structure of the ecosphere are critical to human flourishing? Which relations within and between CASs are critical to the maintenance and even strengthening of their resilience? What behavioural change or technological developments within our power might help to deliver that improved resilience? What social structures are needed to convert these insights into effective policy that will endure over the decadal time scales necessary for them to be effective? It is not my intention in this book to provide answers to these and many other normative questions provoked by this analysis. Rather I wish to open them to further enquiry and also, most importantly, to ask how they can be applied to the research, and perhaps to the development and deployment of geoengineering with its particular challenges of spatial and temporal scale.

Finally, the shift towards systems thinking should be understood as a move beyond rather than away from reductionism. Reductionism continues to be a vital part of the process, but understanding its limitations will allow us to manage climate change in ways more likely to produce useful outcomes and avoid those less likely to do so. Amongst this latter category are, I argue, those with an excessive reliance on prediction. Prediction depends upon pattern forming, where the patterns are based on an analysis of empirical observations of the past and present, combined with rules derived from theories about how the many variables will unfold in the future. An understanding of CAS theory and emergence shows that extrapolating such causal chains into the distant future is fraught with both methodological and epistemological difficulties. Nevertheless, much of the work currently being done around climate change and almost all the work relating to geoengineering, is based on prediction. In the next chapter I examine the extent of this phenomenon and consider its implications for geoengineering policy.

Notes

1 Available online at www.gov.uk/government/uploads/system/uploads/attachment_data/file/287937/07-1184x-tackling-obesities-future-choices-report.pdf, accessed 23 September 2015.

2 Panarchy combines the influences of *Pan* and *anarchy*. Pan is the Greek god of, amongst others things, wild nature and is therefore associated with unpredictability. Anarchy introduces the importance of complex adaptive systems having no exogenous controller and no teleological drivers.

3 Their choice of notation is drawn from traditional sources where α represents a beginning (often faltering); Ω an ending (often in crisis); r is the rate of growth; and K refers to the natural peak achieved by populations given the resources available to them.

4 This assumes a conservative view of our obligations to future generations entailing the preservation of as much as possible of the accumulated capital for the benefit of future generations. Egalitarians might argue that this capital includes too much baggage and that only a radical process of renewal and reassortment will allow the future to emerge free from the environmentally destructive practices of the past.

5 Blue Whale Units were the metric by which whale quotas were measured. Blue whales, at 100 to 150 tonnes each and with a BWU of 1, were the largest whales and the other species were accorded BWUs according to their respective sizes. The target of 16,000 BWUs probably amounted to well in excess of 35,000 whales.

6 http://en.wikipedia.org/wiki/Genocides_in_history, accessed 20 February 2015.

7 The Arab Spring and the First World War were complex political phenomena and it would be a gross oversimplification to suggest that any one incident was their sole cause. However, such incidents can be seen as catalysts that precipitate a particular sequence of events where a kairotic analysis would indicate that a conjuncture making that sequence possible had arisen.

8 The reader will note that Evans' *resilience thinking* is a term I take to be almost synonymous with *systems thinking*. Systems thinking is perhaps a more generic

term. Polasky *et al.* define *resilience thinking* as 'a type of systems thinking that explicitly considers feedbacks, nonlinearities and the sensitivity to change. Resilience thinking places high value on the dynamic processes of learning, adaptation and capacity building' (Polasky *et al.* 2011, 398).

References

Beck, Ulrich. 1992. *Risk Society: Towards a New Modernity*. Sage Publications Ltd.

Beck, Ulrich, and Natan Sznaider. 2006. 'Unpacking Cosmopolitanism for the Social Sciences: A Research Agenda'. *The British Journal of Sociology* 57(1): 1–23. doi:10.1111/j.1468–4446.2006.00091.x.

Capra, Fritjof. 2005. 'Complexity and Life'. *Theory, Culture & Society* 22(5): 33–44. doi:10.1177/0263276405057046.

Clark, Nigel. 2010a. 'Ex-Orbitant Generosity: Gifts of Love in a Cold Cosmos'. *Parallax* 16(1): 80. doi:10.1080/13534640903478809.

Clark, Nigel. 2010b. 'Volatile Worlds, Vulnerable Bodies: Confronting Abrupt Climate Change'. *Theory Culture & Society* 27(2–3): 31–53. doi:10.1177/0263276409356000.

Deschamps, Pierre, Nicolas Durand, Edouard Bard, Bruno Hamelin, Gilbert Camoin, Alexander L. Thomas, Gideon M. Henderson, Jun'ichi Okuno, and Yusuke Yokoyama. 2012. 'Ice-Sheet Collapse and Sea-Level Rise at the Bolling Warming 14,600 Years Ago'. *Nature* 483(7391): 559–564. doi:10.1038/nature10902.

Duit, Anndreas, and Victor Galaz. 2008. 'Governance and Complexity – Emerging Issues for Governance Theory'. *Governance* 21(3): 311–335. doi:10.1111/j.1468–0491.2008.00402.x.

Evans, J. P. 2011. 'Resilience, Ecology and Adaptation in the Experimental City'. *Transactions of the Institute of British Geographers* 36(2): 223–237.

Fuenmayor, R. 1991. 'The Roots of Reductionism: A Counter-Ontoepistemology for a Systems Approach'. *Systemic Practice and Action Research* 4(5): 419–448.

Galaz, Victor. 2012. 'Geo-Engineering, Governance, and Social–ecological Systems: Critical Issues and Joint Research Needs'. *Ecology and Society* 17.

Gunderson, L. H., and C. S. Holling, eds. 2001. *Panarchy: Understanding Transformations in Human and Natural Systems*. Island Press.

Hall, Jim, and Nick Pidgeon. 2010. 'A Systems View of Climate Change'. *Civil Engineering and Environmental Systems* 27(3): 243–253. doi:10.1080/10286608.2010.482659.

Halsnæs, K., P. Shukla, D. Ahuja, G. Akumu, R. Beale, J. Edmonds, Christian Gollier, *et al.* 2007. 'Framing Issues'. In *Climate Change 2007 – Mitigation of Climate Change: Working Group III Contribution to the Fourth Assessment Report of the IPCC*. First edition. Cambridge University Press.

Hardin, Garrett. 1968. 'The Tragedy of the Commons'. *Science* 162(3859): 1243–1248.

Hedaa, Laurids, and Jan-Åke Törnroos. 2001. 'Kairology in Business Networks'. In 17th IMP Conference.

Hedaa, Laurids, and Jan-Åke Törnroos. 2008. 'Understanding Event-Based Business Networks'. *Time & Society* 17(2–3): 319–348. doi:10.1177/0961463X08093427.

Hoffmann, M. J., and J.Riley, Jr. 2002. 'The Science of Political Science: Linearity or Complexity in Designing Social Inquiry'. *New Political Science* 24(2): 303–320.

Holling, C. S. 1973. 'Resilience and Stability of Ecological Systems'. *Annual Review of Ecology and Systematics* 4: 1–23.

Holling, C. 1978. *Adaptive Environmental Assessment and Management*. John Wiley & Sons, London.

Holling, C. 2001. 'Understanding the Complexity of Economic, Ecological, and Social Systems'. *Ecosystems* 4(5): 390–405.

Hornaday, William T. 1889. *The Extermination of the American Bison*. Washington Government Printing Office. www.gutenberg.org/files/17748/17748-h/17748-h.htm.

Hulme, Mike. 2010. 'Cosmopolitan Climates'. *Theory, Culture & Society* 27(2–3): 267.

IPCC. 2008. *Climate Change 2007 – Impacts, Adaptation and Vulnerability: Working Group II Contribution to the Fourth Assessment Report of the IPCC*. First edition. Cambridge University Press.

Kinzig, A. P., P. Ryan, M. Etienne, H. Allison, T. Elmqvist, and Brian H. Walker. 2006. 'Resilience and Regime Shifts: Assessing Cascading Effects'. *Ecology and Society* 11(1): 20.

Lempert, R. J., Steven W. Popper, and Steven C. Bankes. 2003. *Shaping the Next One Hundred Years*. RAND Corporation. www.rand.org/pubs/monograph_reports/MR1626.html.

Levin, Simon A. 1998. 'Ecosystems and the Biosphere as Complex Adaptive Systems'. *Ecosystems* 1(5): 431–436. doi:10.1007/s100219900037.

Levin, Simon A. 2003. 'Complex Adaptive Systems: Exploring the Known, the Unknown and the Unknowable'. *Bulletin of the American Mathematical Society* 40(1): 3–20.

Loorbach, D. 2009. 'Transition Management for Sustainable Development: A Prescriptive, Complexity-Based Governance Framework'. *Governance* 23(1): 161–183.

Loveridge, Denis. 2008. *Foresight: The Art and Science of Anticipating the Future*. New edition. Routledge.

Medd, Will. 2001. 'What Is Complexity Science? Toward an "Ecology of Ignorance"'. *Emergence* 3(1): 43–60.

Miller, C. R. 1992. 'Kairos in the Rhetoric of Science'. In *A Rhetoric of Doing: Essays on Written Discourse in Honor of James L. Kinneavy*, edited by Stephen Witte, Neil Nakadate, and Roger Cherry, 310–327. Southern Illinois University.

Miller, C. R. 1994. 'Opportunity, Opportunism, and Progress: Kairos in the Rhetoric of Technology'. *Argumentation* 8(1): 81–96.

NAS. 1992. *Policy Implications of Greenhouse Warming: Mitigation, Adaptation and the Science Base*. National Academy of Sciences. www.nap.edu/openbook.php?isbn=0309043867.

Park, Robert E. 1928. 'Human Migration and the Marginal Man'. *American Journal of Sociology* 33(6): 881–893.

Pelling, Mark, and Chris High. 2005. 'Understanding Adaptation: What Can Social Capital Offer Assessments of Adaptive Capacity?' *Global Environmental Change* 15 (4): 308–319. doi:10.1016/j.gloenvcha.2005.02.001.

Polasky, S., S. R. Carpenter, C. Folke, and B. Keeler. 2011. 'Decision-Making under Great Uncertainty: Environmental Management in an Era of Global Change'. *Trends in Ecology and Evolution* 26(8): 398.

Richardson, K. A, Paul Cilliers, and M. Lissack. 2001. 'Complexity Science'. *Emergence* 3(2): 6–18.

Ricigliano, Robert, and Diana Chigas. 2011. *Systems Thinking in Conflict Assessment: Concepts and Application*. USAID. http://pdf.usaid.gov/pdf_docs/PNADY737.pdf.

Rittel, Horst W. J., and Melvin M. Webber. 1973. 'Dilemmas in a General Theory of Planning'. *Policy Sciences* 4(2): 155–169.

Rockström, J., W. Steffen, K. Noone, Å. Persson, F. S. Chapin III, E. Lambin, T. M. Lenton, M. Scheffer, C. Folke, and H. Schellnhuber. 2009. 'Planetary Boundaries: Exploring the Safe Operating Space for Humanity'. *Ecology and Society* 14(2): 32.

Schneider, V., and D. Pearce. 2004. 'What Saved the Whales? An Economic Analysis of 20th Century Whaling'. *Biodiversity and Conservation* 13(3): 543–562.

Shove, E. 2010. 'Social Theory and Climate Change'. *Theory, Culture & Society* 27(2–3): 277.

Szerszynski, B. 2010. 'Reading and Writing the Weather'. *Theory, Culture & Society* 27 (2–3): 9–30. doi:10.1177/0263276409361915.

Von Bertalanffy, L. 1972. 'The History and Status of General Systems Theory'. *Academy of Management Journal* 15(4): 407–426.

Walby, Sylvia. 2007. 'Complexity Theory, Systems Theory, and Multiple Intersecting Social Inequalities'. *Philosophy of the Social Sciences* 37(4): 449–470.

Walker, J., and Melinda Cooper. 2011. 'Genealogies of Resilience From Systems Ecology to the Political Economy of Crisis Adaptation'. *Security Dialogue* 42(2): 143–160.

Walker, Warren E., Marjolijn Haasnoot, and Jan H. Kwakkel. 2013. 'Adapt or Perish: A Review of Planning Approaches for Adaptation under Deep Uncertainty'. *Sustainability* 5(3): 955–979.

Wynne, Brian. 2010. 'Strange Weather, Again'. *Theory, Culture & Society* 27(2–3): 289–305. doi:10.1177/0263276410361499.

5 Geoengineering and uncertainty

Most people, and policymakers in particular, have a great aversion to uncertainty. Uncertainty reduces control. Uncertainty increases vulnerability. But too morbid a focus on the negative aspects of uncertainty risks overwhelming its positive potential and can become a source of policy paralysis. This chapter reflects on the nature of uncertainty in relation to climate change and geoengineering. I begin by considering the distinction between risk, uncertainty and surprise, and then examine the aggregation of uncertainties and the irreducibility of uncertainty. The following section looks at the way in which economists deal with uncertainty and finally, I reflect on how the difficulties identified might be better finessed by the application of complex adaptive systems theory in a move away from a fixation on reducing uncertainty towards risk management.

Risk, uncertainty and surprise

Uncertainty is a flexible word. To avoid confusion arising from the conflation of meanings I shall use the words risk, uncertainty, and surprise with specific senses following Knight (1921) and Shackle (1953). Knight distinguished between risk and uncertainty according to whether the likelihood of the event occurring was measurable or not. This usage is followed by many others (e.g. Pryke (2011) and IPCC (Halsnæs *et al.* 2007, sec. 2.3.1)). Accordingly, I use the word *risk* to refer to the likelihood of events for which there is sufficient credible empirical evidence to determine a statistically valid probability distribution function (PDF) for its occurrence. Uncertainty shall refer to events whose possible occurrence is anticipated but for which no credible PDF can be established because of a lack of either suitable empirical data or a viable theory relating causes and effects. Surprise will refer to events whose occurrence is not anticipated, events that were either not conceived prior to their happening, or if conceived, were dismissed as being too remote to be worthy of further consideration. Shackle refers to these as *unexpected* and *counter-expected* respectively (1953, 113).

Climate models represent the relationship between many but not all of the variables at play in the real world, and there will always be some degree of uncertainty attached to the variables not included. This uncertainty is a component of *structural uncertainty*. Mathematical techniques to quantify this uncertainty are undoubtedly helpful in understanding the significance of model outputs but the thrust of the argument about complexity is that, prior to their emergence, the significance of emergent properties is unknowable and therefore incommensurable. These are the Black Swans described by Taleb (2007) and they have the potential of provoking singularities that overwhelm the processes assumed in the models. It may well be that the climate models accurately reflect the contingent futures assumed by their design and given initial conditions, but the techniques of assessing structural uncertainty are themselves limited by their inability to consider the impact of future emergent conditions.

Many authors do not treat risk and uncertainty as distinct. In the space of less than three pages Beck moves from Knight's 1921 distinction between risk and uncertainty, through Keynes and Weber to a twenty-first century of globalised threats and an emerging cosmopolitanism (Beck 2008, 17–18). However, in so doing he somewhat obscures the distinctions between risk, threats and uncertainty. He explains this as the inevitable consequence of the nature of 'late industrial risks' because they defeat any attempt 'to establish yardsticks of risk, such as probability estimates, threshold values and calculations of costs etc.' because of 'the incommensurability of [these] threats and the problem of subjective assessment of probabilities of [their] occurrence' (Beck 2008, 112 quoted from an earlier book by the same author). This analysis does not undermine Knight's original distinction between risk and uncertainty but stresses the need to have rational and empirically supported arguments for the assessment of risk, and that where such arguments are not available the risks become uncertainties. Moreover, it suggests that the distinction between risk and uncertainty, rather than being binary, is a sliding scale because few, if any, rational assessments of risk, even when based on empirical evidence, will be value-free. The subjectivities involved in assessing both risk and uncertainty can plausibly result in one person's risk being another's uncertainty.

The US Bipartisan Policy Center's report on geoengineering illustrates the common usage of the word *risk* in a generic sense to mean *threat* or *danger* whether from commensurable risk or incommensurable uncertainty, referring for example to 'options for addressing climate change risks, regardless of the sources of those risks' (2011, 4). The IPCC definition of *risk* refers to the risk management standard ISO/IEC Guide 73 (2002) that allows a variety of ways of combining probabilities and consequences, one of which is *expected loss*, an economic concept defined as the 'product of probability and loss'. Even beyond the realms of economics, this is a useful way of ranking the acceptability of a risk; a very likely event that causes little harm might be regarded as much more acceptable than a very unlikely event that has catastrophic consequences. The IPCC (Halsnæs *et al.* 2007, sec. 2.3.1) uses *uncertainty* as a generic term synonymous with *likelihood* and provides a confidence scale that factors in the degree and quality of available data, the level of agreement from multiple independent sources about the data, and its significance, indirectly referencing the NUSAP[1] system for assessing uncertainty (Funtowicz and Ravetz 1990, 32).

Surprises are to be distinguished from abrupt non-linear and potentially catastrophic changes, also known as tipping points. A surprise, as I have defined the term, is an event that is not anticipated. Some climate change induced tipping points, for example catastrophic flooding from displacement of the West Antarctic and Greenland ice sheets, or accelerated global warming from methane releases from Arctic permafrost (Lenton *et al.* 2008) are anticipated, even if they are unlikely to happen for many decades or even centuries. The absence of data to support reliable predictions about their timing and scale makes them uncertain but not surprising. Thus, tipping points may be either uncertainties or surprises according to whether the possibility of them occurring has been recognised or not.

An important practical distinction arises from these definitions. It is possible, at least in principle, to formulate specific policies to confront both risks and uncertainties because in both cases they are anticipated, even if there is great uncertainty about how, when or where they will arise. However, this is not so for surprises because, whether unexpected or counter-expected, policymakers cannot knowingly address contingencies of which they are unaware, and do not address those generally dismissed as fanciful.

Conversely, a climate related event cannot be considered a surprise merely by virtue of it occurring in the absence of a pre-planned policy response. The absence of policy may be little more than a failure of governance. Nevertheless, while a surprise-specific policy may not be possible, generic policies designed to build resilience might minimise the impact of surprises, whatever form they take and whenever and wherever they occur. This notion informs much of the ensuing discussion about uncertainty and geoengineering.

Certainty and truth

To avoid confusion in the use of the words *certainty* and *truth* I make the following distinction. Where we are concerned to know how certain subject S is about the truth of proposition *p*, the truth of *p* is a property of *p*, whereas the extent to which S is certain of *p* is a property of S. This allows for *p* to be universally true but for different subjects to be more or less certain of its truth. By this definition, certainty is based on judgement, and in complex systems where the truth may be obscure, even ineffable, S_i may reasonably assess the available evidence about *p* differently from S_j. Indeed, S_i and S_j may have access to different subsets of data about *p* both of which support their different conclusions. Thus, for many centuries many wise and learned men were certain that the Earth was flat and the universe geocentric; the annals of the history of science are replete with the dethroning of similar 'certainties' (Kuhn 1962).

A non-specific sense of uncertainty emerges in the lay public as reported in an NERC funded research geoengineering project *Experiment Earth?* (NERC 2010). Respondents' views about climate change and geoengineering were collected in order to assess their concerns about the many knowledge gaps of which they were made aware. As one is reported as saying:

> I'm still going away with uncertainty and think there's still a lot of uncertainty within the science community and government as to whether these things will work.
>
> (NERC 2010, 77)

His uncertainty seems likely to be of a different order from that of the scientific community. As a lay person, his uncertainty is rooted in his inability to assess the credibility of the scientists, whereas the scientists' uncertainty derives from their ignorance of the science. The respondent's uncertainty is an affect that can only be reduced by reassessing his judgement of the scientists; there is little he can do by way of formalised objective experimentation. Conversely for the scientists, their uncertainty is more likely to yield to such experimentation.

Aggregation of risks, uncertainties and surprises

At the level at which climate change policy is made there are vast amounts of data, analysis, causal chain theorising and prediction, and all manner of extrapolation and interpolation both spatially and temporally, all combined in formal and informal assessments intended to guide policymaking. These inputs will emanate from multiple fields of endeavour, from the most abstract ethical to the most concrete civil engineering considerations, along the way taking in uncertainties in the chemistry, the physics and the social aspects of climate change and the responses to it. The uncertainties in every component of this complex intersection of different realms must be aggregated for policymakers to have some grasp of the likelihood that their chosen policies will deliver their desired policy ends, and how, over the extended periods necessary to make climate change policies effective, this could continue to be so.

Figure 5.1 shows a simple illustration of the effects of aggregating uncertainty. The example projects global GDP growth based on data from the last five decades during which it has averaged about 3.5 per cent per annum with a spread from +6.8 per cent to −2.2 per cent.[2] If these 50 years of historical data are extrapolated a century into the future, the expected outcome is an aggregate growth of slightly more than 30 times. However, this lies within a range of 1 to 100 times. For the actual outcome to be at or close to the expected outcome implies an underlying structural assumption that for the coming

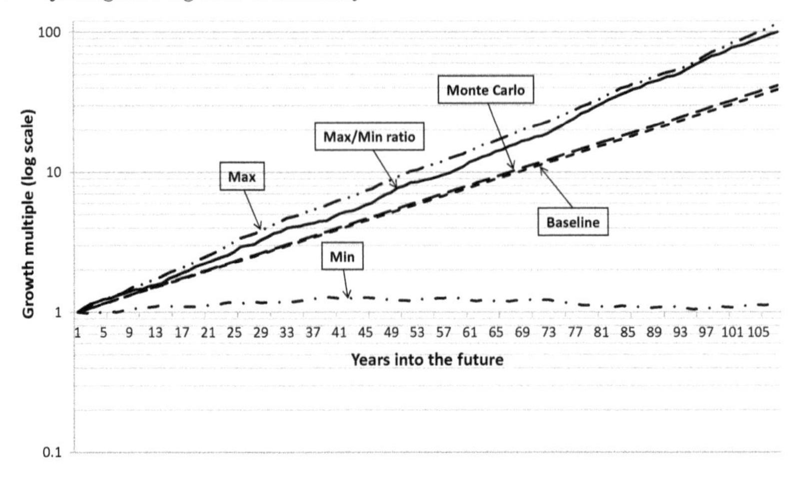

Figure 5.1 Global GDP growth 1960–2011, extrapolated for a century
Source: compiled by the author.

century the world will develop in much the same way as it has in the past 50 years. This may be true, but then again, it may not be.

In reality the mathematics of this example is more complex because each year's GDP growth is not completely independent of the growth in prior years. This dependence is not reducible to a reliable formula as is evidenced by economists' well recognised failure to predict accurately even short-term growth. Over the long term, as Brookfield observed (1990, 69–70) 'It is not meaningful or helpful for either natural or social scientists to speculate about human adaptations a century from now'. There is clearly sufficient historical data to be able to construct a credible PDF of outcomes for global GDP into the distant future but the insurmountable problem is that any policy that assumes that the outcome will be within one or even two standard deviations of the mean, which is the most probable outcome, may well be wholly inappropriate to circumstances where the aggregation results in an outcome in the less likely, but still very plausible, fat tails of the distribution. How much greater is the challenge in predicting all the social and natural phenomena that over decades and centuries, would impact climate change and a geoengineering response to it.

Managing risk and embracing uncertainty

Like shovelling fog out of a ditch, no matter how many uncertainties are resolved, the remaining uncertainty remains much the same, indeed may even be greater, because the continuous emergence of new properties from a panarchy of complex adaptive systems produces new realities with new uncertainties as fast, if not faster, than research can resolve them. It follows that attempts to reduce uncertainty to an acceptably low level, however that is defined, will generally be frustrated. However, while resolving uncertainties does not reduce the remaining uncertainty, it may well reduce risk. Decision makers can act, irrespective of the degree of uncertainty, provided they can control the risks of action. A shift in focus from uncertainty reduction to risk reduction would allow policymakers to open up the possibility of more adventurous experimentation that would promote the heterogeneity that CAS theory holds is vital to ensuring resilience. Policymaking and experiment become one.

The point of empirical research is to find answers to questions as yet unanswered; to bridge knowledge gaps; to resolve uncertaint*ies*; to corroborate or disprove a theory; and so on. Indeed, if there were no uncertainty there would be little point in undertaking the research. However, using uncertainty as a justification to invoke the precautionary principle against research assumes that the foregone knowledge is of insufficient value to warrant the risk of the research. Yet, it cannot be known *a priori* what value the foregone knowledge might have had. There might be genuine concerns about risks associated with the research but the governance of research should be a positive process of pushing gently at the boundaries of uncertainty without taking undue risk, not a negative one of seeking reasons not to do it at all.

There is always the risk that a particular experiment will fail. However, research protocols, when properly devised, will routinely consider the implications of failure and any potential for collateral damage that the research may entail. Health and safety and ethics committees address such questions. Financial and human resource constraints allocate available resources between projects to ensure that the overall programme of research would not be jeopardised by risks emanating from any single project. There are many well tried methods of limiting risk in research programmes. Admittedly these do not eliminate risk altogether, but neither would not doing the research. The paucity of news stories about academic and commercial research activities failing so catastrophically that they threaten the very survival of the institutions sponsoring them, is a testament to their long record of success in avoiding such catastrophes. The evidence is that we are capable of managing research risk and that when failures do occur, they are not only manageable but are also sources of information that help to make further research more productive and less risky.

An heuristic approach to geoengineering would entail a programme of controlled trial and error empirical experiments. The risk taken can be minimised by undertaking experiments on a scale that limits any undesirable effects spatially to small and ecologically less significant locations, and temporally to be quickly buffered by the ecosphere once they are terminated. Those experiments that deliver beneficial results can be selected for enhancement and replication, in a continuing process of gradual increases in scale and intensity; those that do not, can be abandoned before they cause any significant harm. The accumulating empirical evidence would ensure that the up-scaling did not produce increased risk. This heuristic approach to policy embraces uncertainty, reframing it from a threat to an opportunity, and by focussing on risk management, builds on well-established research protocols. The question for policymakers is not what do we need to know to reduce uncertainty to an acceptably low level so that we can act, but rather, given the current level of uncertainty, what is there that we can do that is not unacceptably risky.

As noted, one of the main purposes of empirical experiment is to reduce risk. Risk does not necessarily increase with scale, indeed if it did, few large scale projects could ever have succeeded. The recent NAS report (2015) perpetuates this confusion stating that research should proceed 'from smaller, less risky to larger, more risky projects [...] only as information is collected to quantify the risks at each stage'. While the larger projects might be seen as more risky before the smaller projects are undertaken, the whole point of them is to enable the larger projects to proceed without courting higher risk, indeed, to render the risks of the larger projects even lower than those of the earlier smaller projects.

Economics and uncertainty

Economics is a discipline that has made a major contribution to debates about climate change and geoengineering. Much of this has appeared in the form of cost benefit analyses such as the Stern Review (Stern 2007) and Moreno-Cruz and Keith (2010). In this section I discuss the Ramsey economic models that continue to play a key role in environmental policymaking, and examine the uncertainty implications of discounting.

Many highly regarded economic and integrated assessment models used to assess climate futures, for example the RICE and DICE models of Nordhaus and Solow and the PAGE model used by Stern, are based on the model of national savings originally developed in the 1920s by Ramsey (1928). In their critique of the Stern Review, Beckerman and Hepburn (2007, 191) observe that '[t]he analytical framework most commonly employed to examine questions of intertemporal resource allocation is the tractable workhorse bequeathed to us by Ramsey'. They highlight certain limitations that Ramsey himself identified in a number of simplifying assumptions. Many others considering intergenerational equity have also raised concern about the implications of these abstractions when relying on Ramsey-based models (e.g. Agnani, Gutiérrez, and Iza 2005; Caney 2008; Gerlagh and van der Zwaan 2000; Schelling 1995).

Of particular relevance in relation to climate change are three of Ramsey's assumptions (1928, 543–544). First, he assumes that the community will continue to develop gradually with no major shifts provoked by radical technological developments or population increases, and be driven by the same social dynamics as apply at the outset so that 'enjoyments and sacrifices at different times can be calculated independently and added'. Second, he explicitly ignores 'distributional considerations', assuming that everyone benefits and suffers equally from economic growth and decline. Third, he assumes political and environmental stability 'so that there is no chance of our savings being selfishly consumed by a subsequent generation; and that no misfortunes will occur to sweep away accumulations at any point in the relevant future'.

It need not be laboured that over the timescales at play in geoengineering and with the possibility, if not the probability, of climate change events that may literally 'sweep away accumulations at any point in the relevant future', when simulating long-term climate futures using an economic model constrained by any one of these abstractions serious doubts could be raised about the soundness of its outputs, and when taken together they severely undermine the arguments of any economist relying upon them without examining their implications.

In different ways all three assumptions impinge on the setting of the discount rate to be used in assessing policy alternatives. Before addressing these issues, a brief introduction to discount rates will be helpful. Discounts rates are a mathematical device to enable cash flows that arise at different times to be aggregated and compared. In its simplest formulation, an investor who offers to invest £100 provided he gets back at least £105 after one year, is discounting the future benefit at 5 per cent per annum. This is a measure of what economists refer to as his time preference. This concept has been extended to compare a preference between consumption today and consumption in the future; in effect measuring non-specific but quantified personal expenditure across time. However, one's preference to spend £30,000 on a car today or £38,000 on a holiday in 5 years, or £49,000 on a home in 10 years, or £130,000 for residential and nursing care in 30 years, are not obviously comparable merely because they are all reduced to equal monetary values when discounted at 5 per cent per annum. Moreover, people from different cultural settings are quite likely to value these expenditures differently implying that everyone has their own personal time preference rather than there being 'a' time preference that is appropriate for all people for all forms of costs and benefits.

The UK Treasury's Green Book[3] presents their guidelines for appraising national-scale infrastructure investments. A central part of the process requires future cash flows to be discounted in order to compare the utility of alternative projects across different periods of time, and even across different generations. They propose the Social Time Preference Rate (STPR) as 'the standard real discount rate'. The STPR has four parameters, δ, L, μ and g, representing respectively, an individual's pure time preference, the risk of there being a catastrophic failure that overwhelms the expected outcome for that individual, the elasticity of the marginal utility of consumption to accommodate the notion that a fixed amount of expenditure produces less value the wealthier one is, and annual growth in per capita GDP. Based on the literature, the UK Treasury guidelines arrive at a unified STPR of 3.5 per cent. This compares with Nordhaus' 6 per cent and Stern's 0.1 per cent. The effect of this choice is dramatic. Whereas Nordhaus, using a 6 per cent p.a. discount rate, would calculate the present value of a cost or benefit of £1bn arising in 100 years' time to be only £3m, Stern using a rate of 0.1 per cent would value it at £900m, with the Green Book's STPR at £32m. These huge differences completely transform political assessments based on cost benefit analysis, greatly preferring the interests of more proximate generations against the more distant, or vice versa according to whether the discount rate is higher or lower. Since the balancing of these interests is a normative and ethical matter, justifications for which rate to use, or indeed whether the methodology is even appropriate in the case of long-term global intergenerational distributions, must lie outside the realms of economics.

Where the investment is being appraised over longer than 30 years, the Green Book proposes an exponentially descending series of discount rates. In the example in the previous paragraph, this would almost triple the Green Book's valuation from £32m to £84m, but still produce a figure less than a tenth of Stern's. This procedure is justified by reference to, amongst others, Weitzman (2001; 1998) and corroborated by reference to Gollier (2001). Interestingly, Weitzman uses this method of appraisal specifically to assess proposed investments in response to global warming; by extension one must assume that he would also consider it to be applicable to investments in geoengineering.

The Green Book's silence on the abstractions made by Weitzman and Gollier in their analyses is a matter for concern. Weitzman limits his analysis to 'one-time irreversible investment decisions'. However, investments responding to climate change are made continually year on year into the future and it always remains possible for them to be revisited and revised. Gollier's contribution is based on a simplified 'one-good, pure exchange economy with identical consumers' (Lucas 1978). Each of these identical consumers is a 'paternalistic agent [who] wants to maximize the net present value of the flow of future expected utility' (Gollier 2002). The Green Book simply ignores the structural differences between these highly abstracted economies and the real world.

A number of concerns arise when using discount rates in long term infrastructure investment appraisals. First, they are built on extreme abstractions from reality and there is little or no analysis of how relaxing these conditions would impact the conclusions. It is much as if those attempting to solve the longitude problem had declared that because a pendulum clock works on land, we can argue that it will work at sea by the simple expedient of assuming that the ocean is always calm and flat and the ships never pitch or roll in the wind. Gollier does relax his initial assumption that the future is recession-proof, but he does not consider the consequences of multiple goods, or non-maximising non-paternalistic agents each with his own set of local values. Weitzman does acknowledge that his assumption that investment decisions can never be revisited does require further thought but he defers it for consideration elsewhere, and if that work was ever done, it does not inform the advice offered and followed by the UK Treasury.

Additionally, the economies on which the economists' arguments are constructed are ones where there is national sovereignty, a condition that does not exist in the anarchic global panoply of interdependent states. Equally abstracted is the notion that each sovereign state comprises a homogeneous population as regards time preference whether with regard to personal savings or major national, even global, infrastructure investments to combat climate change. This being so, it is difficult to see how any discount rate could be agreed upon that would adequately reflect the interests of the global population today and also apply continuously into the distant future, even if this were achieved by some form of reducing rate.

A further concern turns on three abstractions hidden within this methodology. The first is that individuals' time preference can be aggregated to arrive at a community's combined time preference, an abstraction also made by Ramsey. For aggregation to be meaningful requires that the entities being aggregated are sufficiently alike in all material respects. As already noted, one individual may have different time preferences and attitudes to risk according to both her personal circumstances and the nature of the proposed expenditure and its supposed benefits. The time preferences may also vary through time as her circumstances and her values and perspectives mature on her journey through the 'seven ages' of her life. Individuals in different cultural settings, whether distinguished geographically or temporally, are likely to have widely different time preferences, and indeed, value life itself very differently. Aggregating time preferences certainly makes the mathematics more tractable, but in so doing it increasingly divorces the economic theories from reality. Indeed, Rittel and Webber were explicit that there is no welfare function that captures a unified sense of utility for any major social investment project (Rittel and Webber 1973).

The second hidden abstraction is that an individual's time preference is the same whether she is both the investor and the beneficiary of whatever the investment produces, or, as will generally be the case in long-term climate change projects, she is the investor but the beneficiary is someone else, in a different place at a different time, a person with whom she has little or no connection other than both being members of the human race, although probably not at the same time. The economists' assumption that an individual's time preference will be the same even when others benefit from her investment, may well be true for a small number of extreme egalitarians, but is most unlikely to be true for the vast majority of people. To some extent this is dealt with by reducing discount rates for longer periods, but even this does not eliminate specious accuracy in valuing competing policy options.

The third assumption is that preferences can be monetised and that their values are equally applicable not only across generations but also between cultures. The value of retaining the rain forests to support its indigenous wild life and people is simply incommensurable with a desire for more teak furniture in the homes of the affluent or increased soya bean production as cattle feed for environmentally demanding meat protein.

In all three cases, the abstractions are necessary to accommodate the mathematical rigours of a reductionist analysis. But as these abstractions show, this approach fails to recognise that the systems in which the investments are occurring are complex adaptive systems with emergent properties that will generally, and in the long-term almost always, confound prognostications based on the underlying assumption that we know what the world is like and that things will unfold more or less following a smooth extrapolation in line with the established patterns from the observable past, through the present into the immediate and then the distant future.

In a testament to his status as a world-class academic economist, Weitzman was able to conduct an unusually large Delphic sample of more than 2,000 internationally diverse PhD level economists who set the STPR applicable to investments to mitigate the effects of global warming at various levels between -3 per cent and +27 per cent per annum. From this data he concludes first, that the Delphic responses corroborate his theory justifying a descending rate of STPR as the appraisal period extends into the distant future; and second, that the appropriate absolute values are on an exponentially descending scale from 4 per cent to 0 per cent as the period lengthens from the immediate to more than 300 years. His table of descending discount rates is only marginally different from that recommended in the Green Book.

Weitzman's use of Delphic opinion is certainly interesting but it is does not constitute proof that the generality of present and future individuals share the values and aspirations of 2,000 carefully selected experts each of whom, despite their varied cultural backgrounds, has become deeply absorbed into the elite academic culture of 21st century Western market-based economics. Indeed, Weitzman amusingly notes that 'an "expert" here might be defined as an economist who knows the literature well enough to be able to justify any reasonable social discount rate by some internally consistent story' (2001). The general consistency of their responses is likely to be more a manifestation of their education and academic engagement, than an objective statement about a generally applicable discount rate. In effect, Weitzman has turned the entire notion of time preference on its head. Rather than looking for evidence of time preference in the expressed behaviour of individuals, Weitzman has obtained a 'professional' view of its numeric value and ascribed this to the wider community. There is a need to demonstrate that the experts' views conform to those of their wider local communities.

We know that the indefinite future will not be inhabited by either infinitely-lived agents with unchanging tastes and motivations nor by people enjoying an unending libertarian future characterised by continuous economic growth delivered within a broadly stable neoclassical capitalist economic environment. Even Nordhaus, the author of the DICE model, recognises this difficulty in his 1992 observation that only 80 years previously:

> the Ottoman, Austro-Hungarian, British, and Russian empires ruled much of the world. In America there were no income taxes, women

> could not vote, each person commanded 1.5 horsepower, and the major polluters were the 21 million horses.
>
> (Committee on Science, Engineering, and Public Policy 1992, 516)

Given the lively debates even today between neo-Keynsians and neo-Hayekians (Mason 2011) whose irreconcilable positions on economic management demonstrate the challenges that any economic theory has in predicting even relatively short-term outcomes within a single nation-state, it is unsurprising that economists struggle to provide any consensus about the long-term global economic outcome from different contingent climate futures. Van Vuuren *et al.* (2011) compare a number of climate models and conclude that the differences between them potentially represent trillions of dollars of cumulative abatement costs, on the basis of which they counsel extreme caution on the part of policymakers intending to rely on them.

Indeed, it may be that economics as a discipline is bereft of tools fit for purpose to assess long-term climate futures. Considerable caution is necessary in the use of economic models until a) there is more clarity about the setting of discount rates; b) the challenges posed by Ramsey's modelling assumptions that introduce radical structural uncertainty into long-term modelling are addressed; and c) there is a reconciliation of the inconsistencies introduced by attempts to monetise all the costs and benefits and treat them all as if they were commensurable, ignoring many uncertainties and surprises that are demonstrably not. Perhaps economists should take heed of Jamieson's (1996) caution to climate scientists that they should show some humility with their projections. DeCanio is one of a number of economists who share this view. He observes that the economists':

> ability to solve mathematical puzzles ... conveys a sense of power and accomplishment that is easily mistaken for *real wisdom*.
>
> (DeCanio 2003, 158 emphasis in original)

Prediction

It is in the nature of climate change policy formulation that much effort is invested in contemplating potential future climate states and their social and wider ecosystem implications. Charlesworth and Okereke (2010) note that '[c]limate prediction and projections have become without a doubt the bedrock of climate policy at least in the developed world' (2010, sec. 2.1). Whatever methods are used to predict the future, whether the Delphi method that relies on expert opinion, or sophisticated AOGCMs, the predictions are inescapably subject to uncertainty. How this uncertainty is assessed and managed lies at the heart of the challenge for geoengineering policymakers.

Sarewitz *et al.* (2000) argue that prediction serves one of two purposes: explanation and decision making. This distinction parallels Jonas *et al.*'s (2014) distinction between diagnostic (retrospective) and prognostic (prospective) uncertainty. Explanatory predictions seek to corroborate (or refute) a prior hypothesis, whereas decision making predictions are concerned with bringing about (or avoiding) certain future states. Explanatory prediction is an activity that falls squarely in the domain of Kuhn's normal science. It is also the means by which Popperian falsification succeeds in rejecting theories when the empirical evidence does not match a theory's predictions (Kuhn 1962). Decision making predictions, by contrast, create the future, whether they are adopted into policy or not, because they limit the available policy choices and so reflexively constrain the volatile bifurcations at the heart of the path dependency that determines the future. They cannot be used to corroborate a prior theory about the future because they reflexively alter the context in which the future unfolds, rendering causal chains increasingly opaque the greater the temporal and spatial horizon.

Using ten case studies from a range of man-made and natural disasters, Sarewitz *et al.* illustrate the value and limitations of prediction as a policy tool and derive criteria by which its appropriateness to any social policy arena might be assessed. As they note, prediction is at the heart of the reductionist approach to science but only recently has it been extended as part of the scientisation[4] of policymaking not only to improve policy but also 'to reduce political risk' for the politicians (2000, 17). Today's decision makers, like their predecessors back to and beyond the biblical Noah, are faced with the challenge of making judgements about an inescapably uncertain future; the major difference is the quantity and quality of the evidence available to them. Noah's belief in God has been replaced by an equally powerful belief in the potential of science to predict alternative futures. But Sarewitz *et al.* caution that that 'can become a substitute for political and moral discourse' (2000, 17). For example, they identify a mutually convenient arrangement between policymakers and scientists in which the scientists secure funding for their favoured projects enabling the policymakers to 'point to predictive research as "action" [...] while deferring difficult decisions as they await the results of research' (2000, 18). By this means, a deep political aversion to risk has resulted in science becoming increasingly about justifying inaction.

Sarewitz *et al.* (2000, 18–20) set out criteria by which to evaluate the policy value of predictive research: a) *timeliness* – are useful predictions likely to take more time to develop than policymakers have to react to the underlying issue? b) *necessity* – are accurate predictions necessary for decision makers or do they have viable alternatives? c) *accessibility* – can the predictions be readily interpreted by non-experts to avoid their misuse or abuse? d) *compatibility* – are the criteria for success in scientific prediction sufficiently similar to those that apply in policymaking to avoid confusion and controversy? e) *distraction* – is there a risk that a focus on prediction will drain financial and other resources away from other activities that might better address the issue at hand? In the next chapter, I apply these criteria to geoengineering.

Returning to the distinction between explanatory and decision making predictions, Sarewitz *et al.* summarise the problematic of prediction as a policymaking tool by observing that:

> it requires an extrapolation of the concept of scientific prediction itself, from its traditional significance as a test of fundamental and reductionist laws of nature, to a newer role as a technique that seeks to extract policy-relevant predictive certainty from research on *complex processes.* Given the difficulties of achieving such relevant certainty, the role of scientific prediction in policy making is itself highly uncertain.
>
> (2000, 21, emphasis added)

Charlesworth and Okereke (2010, sec. 2.1) also consider prediction to be of limited value, noting that 'there may be considerable risk attached to policy processes that simply assume uncertainties [of the Earth System] can be removed'. Sarewitz *et al.* (2000, 380) conclude that prediction is least appropriate where there is: a) low or unknown predictive skill; b) little experience in using predictions with the phenomena in question; c) long characteristic time (the period over which predicted events occur); d) at least one alternative to prediction available; and e) high uncertainty in outcomes of alternative decisions.

The centrality of predictability in determining policy towards geoengineering is discussed by Jamieson (1996). Referring to conclusions from cognitive psychology he notes 'that people tend to be overconfident about their judgments' and suggests that '[t]his result should give experts about geoengineering some humility about the reliability of their predictions. Their guesses about what will occur may be no better than those of novices'. He continues:

> If we couple the pervasiveness of unintended effects and the tendency of experts to overestimate their expertise with the incredible complexity of the climate system, the grounds for skepticism about reliably predicting the effects of [geoengineering] seem very strong. Not only is there reason to doubt that the consequences of [geoengineering] can be predicted reliably, but there is reason to be suspicious of those who claim otherwise. It might be claimed that this condition rules out too much: reliable prediction is hard to come by. I agree, but reliable prediction has been one of the central goals of science since at least the seventeenth century, as well as one of the stated goals of the US Global Change Research. If technological interventions have the potential to bring about quite profound negative effects, it is not too much to ask

> that their advocates know what they are doing. And in this case, there is little evidence that they do.
>
> (Jamieson 1996, 327)

In this extract Jamieson illustrates the dominance of reductionist thinking. In conventional linear fashion, he argues that scientists have a responsibility to 'know what they are doing' when their interventions may have 'quite profound negative effects' and therefore they should deliver more reliable predictions in order to minimise those effects, or possibly avoid them altogether. However when confronted by intractable uncertainty and a tendency to overstate their claims, a more consistent conclusion would be to expect the advocates of geoengineering neither to know with certainty what they are doing, nor to be capable of delivering reliable predictions, and therefore to develop policy accordingly. Jamieson is setting up an unachievable objective because he fails to make the distinction between explanatory and decision making prediction.

The common practice of recognising irreducible uncertainty yet nevertheless maintaining that prediction can still be sufficiently robust to inform policy is discussed by Charlesworth and Okereke (2010). They cite (2010, sec. 2.1.1) an IPCC FAQ concerning the likelihood of abrupt climate change and paraphrase the IPCC response as 'not very likely, we cannot dismiss the possibility but we do not really know'. They argue that the tension between 'not very likely' and 'we do not really know' is significant and explore this with further quotations from the IPCC. They also identify, from a broad range of authorities, the questionable proposition that because critical thresholds in the climate system can be imagined, they can therefore be predicted and controlled. However, they note that there is invariably a paucity of detail as to how that imagination is to be operationalised or how we are to know that we have imagined all the possible critical thresholds.

They suggest that our historical and continuing failure to predict and control critical thresholds both in nature and society does not give cause for much optimism that climate change will be any different. As examples they cite avalanches, anaphylactic shock, heart attacks, multiple organ failure, volcanoes, domino effects, economic crashes and political revolutions. In all these cases their happening in the future is taken for granted; they are not surprises notwithstanding the considerable uncertainty surrounding their timing, their severity and possibly their location. Yet this kind of uncertainty does not preclude anticipatory policy responses. A prime example is the specification of earthquake resistant building design in zones prone to tremors. A major earthquake in London would be a surprise but the flooding of the Thames would not, which is why the Thames Barrier was built but London buildings, unlike those in Tokyo and Los Angeles, are not designed to withstand major earthquakes.[5] Self-evidently, examples of contingent events we have yet to conceive, Shackle's *unexpected* surprises, cannot be cited but

Charlesworth and Okereke dismiss the prospect of a world without surprises. As they rather graphically summarise:

> at present, there is no robust way of finding all critical thresholds in global and more local ecological systems without taking the risk of blowing up the 'laboratory' along with the experimenters.
>
> (Charlesworth and Okereke 2010, 124–125)

It is in the nature of surprises that we cannot assess the extent of the structural uncertainty they cause in our models. As Brunner explains, models of complex adaptive systems will always be unreliable because 'the context and rules of interaction of the agents in [such a] system are always open to new experience and insight and the system never reaches equilibrium' (Brunner 2000, 300). The paradox is that even if more work did result in climate models producing totally accurate predictions, that fact could not be known until it is empirically tested by which time the future will have become the present and the opportunity to influence that present will have been irretrievably lost. It seems that Cassandra's fate is shared by her twenty-first century acolytes.[6]

This is not to argue that model predictions have no value, indeed they perform a vital role that in many ways is made more valuable by the wide spread between them so starkly reminding us of the extent of our ignorance. Their importance is in the qualitative rather than quantitative understanding of the interaction between variables. In later chapters I will return to the relationship between long-term climate model predictions and near-term climate policy formulation.

CAS theory, uncertainty and geoengineering

I now turn to consider whether, with the extended spatial and temporal scales of geoengineering, CAS theory might be better at confronting the challenges provoked by uncertainty than the dominant reductionist predict and control methodology. I shall highlight two aspects of Holling's work on complex adaptive systems that relate directly to uncertainty, its unavoidability and the need to approach it heuristically.

Holling recognises that uncertainty is unavoidable and looks for ways to accommodate rather than resolve it, as he puts it:

> Embrace uncertainty and unpredictability. Surprise and structural change are inevitable in systems of people and nature.
>
> (Holling 2001, 391)

Indeed, he argues that it is precisely when systems are at their most vulnerable that uncertainty and surprise can be the source of the innovation needed for future growth and flourishing.

Second, the model of enquiry into socio-ecological systems that he sets out is heuristic and therefore embraces failure as part of the learning process. He makes a key distinction between fail-safe and safe-fail. In the former the intervention is designed not to fail because a failure could be catastrophic – a nuclear reactor or aircraft are prime examples of fail-safe design. In these cases, the usual consequence of a failure is a catastrophe – Chernobyl, Fukushima, Bhopal, and thalidomide, to name just four. Safe-fail design, on the other hand, assumes that the interventions will on occasions fail but designs them so that the failures are not catastrophic. In the dominant discourse of geoengineering we see only fail-safe approaches. The predict and control mindset seeks to remove uncertainty and while it may well use failures as learning opportunities, it does not, except in controlled laboratory environments, instigate policy initiatives expecting some of them to fail. Politicians are rarely heard to say, when introducing new policies for defence, education, health, taxation or any other area of public policy, 'let's try this, we're not sure if it'll work, but if it doesn't, the damage shouldn't be too great and we might learn something that will help us do better next time'.

Practices that attempt to eliminate the risks and uncertainties are, by Holling's arguments, more likely to stifle the creativity that is needed to determine whether an intervention has value and if so in what way, than they are to promote it. Much as Ravetz observed, solutions will not emerge from laboratory science fully fledged and ready to go but rather they will evolve through an iterative process of success, failure and learning, consistent with Holling's heuristic adaptive management. The challenge in such an approach is in ensuring that the consequences of failure are not only controlled and capitalised upon, but equally importantly, not pounced upon by the media as reasons to vilify the scientists and policymakers and in so doing stifle creativity.

The impact of geoengineering on ecological and human social resilience is unresolved. In the geoengineering literature resilience is generally seen as an outcome of adaptation rather than of emissions abatement or geoengineering (e.g. Bipartisan Policy Center Task Force on Climate Remediation Research 2011; Bodansky 1995; Caldeira 2009; Carlarne 2011; Fox 2009; Mabey 2011; MacCracken 2009a; Ott *et al.* 2004; Malone and Rayner 1998). Indeed, consistent with my earlier observations concerning palliative and curative responses to global warming, MacCracken (2009a, sec. 5.3) considers that while adaptation and increasing resilience address the symptoms of climate change, attempting to reduce their negative impacts, geoengineering seeks to address its root causes and thereby reduce the intensity of its impacts. However, elsewhere he argues that geoengineering might be 'a complement to adaptation and the building of resilience' when it is deployed as a means of avoiding threats to socio-ecological resilience from catastrophic climate change, that might otherwise have required even more heroic remedies with potentially more serious risk profiles (MacCracken 2009b, 11).

Others are less accommodating, regarding the uncertainty as having only negative potential. Luke, a political scientist, regards all forms of geoengineering as inimical to ecological and social resilience because in place of 'resilient, micro-scale and reversible solutions' what they offer is the 'vain hope' of rapid solutions that would be 'singular, brittle, macro-scale, and possibly irreversible' (Luke 2010, 123). As already noted and will be examined in more detail in Chapter 7, these are largely ideological arguments because they make assumptions about as yet unknown knowledge and wrongly frame geoengineering as a dominant response to climate change.

CAS theory argues that the presence of a global controller undermines the resilience of the system because the pursuit of *its* ends must threaten heterogeneity by virtue of them being privileged over those of others. The greater the diversity within the system the greater the prospects for the adaptations necessary for its resilience. As Darwin noted 'better adaptation in however slight a degree to the surrounding physical conditions will turn the balance' (1859, chap. 14). Even if an altruistic global controller aspires to promote adaptability, it has two challenges. First, it cannot guarantee that it will get it right every time and because it has the power to operate at system-wide level (being a *global* controller) one of the surprises may well be that one of those errors could be fatal for the system as a whole. Second, as noted above, a human global controller (whether an individual or an institution) will overwhelm the interests of other humans as well as other species with a resulting loss of diversity as is already evidenced across the planet. Even though localised actions may also have bad, even disastrous, outcomes, because they are local they pose much less threat to the entire system. Their collapse may even release resources to adjacent CASs propelling them more securely along their adaptive cycle.

The ecosphere is a CAS that comprises an uncountable number of subsystems. At the level of any of those sub-systems a global controller may well cause its destruction without unduly affecting the stability of the ecosphere as a whole because other systems within the ecosphere will act to accommodate the lost system. However, if the global controller is operating at the level of the ecosphere itself, the risk of failure assumes an altogether different order of magnitude. Geoengineering intervenes at or close to the level of the ecosphere.

In principle, it seems therefore that humans are faced with a dilemma; when acting as a global controller of the climate system we cannot impose order on the system as a whole in such a way as to serve our needs without simultaneously undermining the heterogeneity upon which those needs depend. The argument here is not that human intervention will have this, that or the other effect, but rather that global interventions risk undermining the resilience of the CAS with potentially system-wide catastrophic consequences. No prediction is necessary. We do not need to know precisely how or when the CAS would collapse from the behaviour of a global controller, it is sufficient to know that a collapse would be the inevitable consequence even if the timing and the nature of the collapse must remain a matter of speculation. A parallel point is well made when summarising Hayek's views on what he considers to be the unfulfillable quest at a national scale for economic stability, Walker and Cooper comment:

> The natural complexity of market phenomena was such that no centralized authority could hope to predict, much less control, the precise evolution of individual elements in the system. At worst, such efforts risked inducing long-term crises that would not have occurred without the undue interference of the state.
>
> (Walker and Cooper 2011, 149)

Discharging the role of global controller implies intention. If a regulatory arrangement is established, whether at national, regional or global scale, this will demand the articulation of normative criteria for desired climate outcomes if the minutiae of regulation are to be codified, disseminated and used as a basis for behavioural change. We know that different components of the climate system operate on vastly different time scales, from the more or less instantaneous behaviour of the weather and its immediate impact on life, to processes that unfold over geologic time but are equally vital to the climate system. If human intention were to be capable of preserving the far-from-equilibrium stability of the climate system it would need to attend not just to those processes that affect us in the here and now and perhaps the next few decades, but equally importantly, to those that will play out over centuries and even millennia. Do we have, indeed could we conceivably have, enough knowledge to reduce uncertainty about the consequences of our actions over these extended time scales to 'acceptably low' levels, whatever that might mean, where we could intervene with the confidence that our actions would not threaten the integrity of the entire system? Or would we be merely clicking up another notch in the ratchet of Beck's risk society, creating problems for the distant future in the Panglossian hope that our successors will have the human and material capital, and the time, to confront them successfully as argued by commentators such as Lomborg (2007) and Beckerman (2004; 2008).

In summary, looking at geoengineering through the lens of CAS theory suggests some boundaries within which it might operate to minimise threats to the resilience of the systems on which humanity depends: multiple technologies, localised modest scale deployment, buying time for abatement and adaptation to bite. These and other measures whose purpose is to promote heterogeneity in the technology, in the responses, in the natural world and in human society, begin to emerge as the prudent options. They allow for, indeed encourage, errors and corrections and promote adaptability as circumstances unfold in unexpected ways. Uncertainty ceases to be a source of policy paralysis and becomes a driver of policy innovation. The objective is not to minimise or even eliminate uncertainty but rather by embracing it and focussing on risk management, to make it work for us.

Prudence requires us to be drawn towards uncertainty for it is there that the future lies.

Conclusion

Geoengineering is a child of the predict and control scientific method that emerged from the dominant normal science paradigm (Kuhn 1962) of the Enlightenment and in which uncertainty is theorised as epistemological. As the twentieth century closed, this paradigm came under increasing pressure as complex adaptive systems with their emergent properties and their ability to regenerate and renew even in the face of major perturbation were distinguished from complicated but deterministic systems. But the argument presented here suggests that the geoengineering discourse has yet to manifest a transition to this new paradigm. This is evidenced by the overwhelming reliance on expert-analytic methodologies predicated upon the Baconian idea that we can dominate nature by more research that will lead to more knowledge that will reduce uncertainty to some 'acceptably low' point. Yet, despite this framing, in Chapter 7 we will encounter early tentative signs of systems thinking within the geoengineering literature.

By approaching geoengineering through its impact on resilience we can look to Holling for inspiration. He closed an early paper on the resilience and stability of ecosystems by observing that from a management approach based on resilience would flow:

> not the presumption of sufficient knowledge, but the recognition of our ignorance; not the assumption that future events are expected, but that they will be unexpected. The resilience framework can accommodate this shift of perspective, for it does not require a precise capacity to predict the future, but only a qualitative capacity to devise systems that can absorb and accommodate future events in whatever unexpected form they may take.
>
> (Holling 1973, 21)

The reframing of geoengineering as part of a package of responses to climate change that is understood from a complexity perspective would have important policy consequences by releasing many of the tensions that currently inhibit progress. These and related issues are the subject of Chapter 7 but first I examine the current balance between predict and control and systems thinking in the public policy.

Notes

1 The NUSAP scheme for expressing uncertainty requires an assessment of each of the following attributes of any datum in order to determine the degree of its uncertainty. Uncertainty is usually expressed in terms of the error bars given by

S(pread) but this numerical value is to be mediated by an assessment of the quality of this spread as determined by A and P where N = numeral or value that might not be a number; U = unit e.g. £, g, clicks, etc.; S = spread or range of values e.g. ±n or ±n%; A = assessment that records the (un)reliability of the quantitative information in the earlier categories e.g. confidence limits or 'high' or 'low'; and P = pedigree, a complex system for assessing the provenance of the data.

2 Data from the World Bank databank, available online at www.worldbank.org (accessed 7 August 2012). The maximum projection is calculated by assuming that the growth in every year will be a random percentage between the sample median and maximum. The minimum projection is calculated by assuming annual growth of a random percentage between the sample median and minimum. The Monte Carlo simulation is based on 1,000 iterations in which the annual growth is selected randomly from within the sample range and, as expected, more or less coincides with the baseline projection assuming continuous growth at the sample mean.

3 The Green Book is available from the UK government website at www.gov.uk/government/uploads/system/uploads/attachment_data/file/220541/green_book_complete.pdf (accessed 30 November 2014).

4 *Scientisation* refers to the process whereby science has become an increasingly important factor in policymaking both because, in a Beckian sense, it contributes to the creation of a range of social ills, and also because it contributes to their solution (Eden 1996, 188).

5 UK building regulations do not generally take account of seismic risks – BS EN 1998 and PD 6698.

6 Gifted by Apollo with the power of prophecy Cassandra foretold the fall of Troy, the ruse of the Trojan horse and the death of Agamemnon. However, as a punishment for refusing to become Apollo's consort, he also procured that no one would believe her warnings. Of course, had he not done so and had her warnings been heeded, the prophesies would have been rendered untrue.

References

Agnani, B., M. J. Gutiérrez, and A. Iza. 2005. 'Growth in Overlapping Generation Economies with Non-Renewable Resources'. *Journal of Environmental Economics and Management* 50(2): 387–407.

Beckerman, Wilfred. 2004. 'Intergenerational Justice'. *Intergenerational Justice Review*. www.intergenerationaljustice.org/images/stories/publications/gg12_20040629.pdf.

Beckerman, Wilfred. 2008. 'The Impossibility of Intergenerational Justice'. *Intergenerational Justice Review*.

Beckerman, Wilfred, and Cameron Hepburn. 2007. 'Ethics of the Discount Rate in the Stern Review on the Economics of Climate Change'. *World Economics* 8(1): 187–210.

Beck, Ulrich. 2008. *World at Risk*. First Edition. Polity Press.

Bipartisan Policy Center Task Force on Climate Remediation Research. 2011. *Geoengineering: A National Strategic Plan for Research on the Potential Effectiveness, Feasibility, and Consequences of Climate Remediation Technologies*. Bipartisan Policy Center. http://bipartisanpolicy.org/sites/default/files/BPC%20Climate%20Remediation%20Final%20Report.pdf.

Bodansky, Daniel. 1995. 'The Emerging Climate Change Regime'. *Annual Review of Energy and the Environment* 20(1): 425–461.

Brookfield, Harold. 1990. 'Vulnerable Places; Vulnerable People; Human Science Approaches to Problems of Adaptation'. In *Global Change: The Human Dimension*. Academy of the Social Sciences in Australia.

Brunner, R. 2000. 'Alternatives to Prediction'. In *Prediction: Science, Decision Making, and the Future of Nature*. Island Press, 299–313.

Caldeira, Ken. 2009. *Geoengineering: Assessing the Implications of Large-Scale Climate Intervention*. Washington, DC: US House of Representatives Committee on Science and Technology. http://democrats.science.house.gov/Media/file/Commdocs/hearings/2009/Full/5nov/Caldeira_Testimony.pdf.

Caney, Simon. 2008. 'Human Rights, Climate Change, and Discounting'. *Environmental Politics* 17(4): 536. doi:10.1080/09644010802193401.

Carlarne, C. P. 2011. 'Arctic Dreams and Geoengineering Wishes: The Collateral Damage of Climate Change'. *Columbia Journal of Transnational Law* 49: 602–755.

Charlesworth, Mark, and Chukwumerije Okereke. 2010. 'Policy Responses to Rapid Climate Change: An Epistemological Critique of Dominant Approaches'. *Global Environmental Change* 20(1): 121–129. doi:10.1016/j.gloenvcha.2009.09.001.

Committee on Science, Engineering, and Public Policy (US). 1992. *Policy Implications of Greenhouse Warming: Mitigation, Adaptation, and the Science Base*. The National Academies Press. www.nap.edu/catalog.php?record_id=1605.

Darwin, Charles. 1859. *On the Origin of Species by Means of Natural Selection, or the Preservation of Favoured Races in the Struggle for Life*. John Murray.

DeCanio, S. J. 2003. *Economic Models of Climate Change: A Critique*. Illustrated edition. Palgrave Macmillan.

Eden, Sally. 1996. 'Public Participation in Environmental Policy: Considering Scientific, Counter-Scientific and Non-Scientific Contributions'. *Public Understanding of Science* 5(3): 183–204. doi:10.1088/0963–6625/5/3/001.

Fox, Tim. 2009. *Climate Change: Have We Lost the Battle?*. Institution of Mechanical Engineers. www.imeche.org/NR/rdonlyres/77CDE5E4-CE41-4F2C-A706-A630569EE486/0/IMechE_MAG_Report.PDF.

Funtowicz, Silvio, and Jerome R. Ravetz. 1990. *Uncertainty and Quality in Science for Policy*. Springer.

Gerlagh, R., and B. C. C. van der Zwaan. 2000. 'Overlapping Generations versus Infinitely-Lived Agent: The Case of Global Warming'. In *The Long-Term Economics of Climate Change*, edited by Richard Howarth and Darwin Hall, 3: 301–327. Stamford: JAI Press.

Gollier, Christian. 2001. 'Should We Beware of the Precautionary Principle?' *Economic Policy* 16(33): 301–328.

Gollier, Christian 2002. 'Time Horizon and the Discount Rate'. *Journal of Economic Theory* 107(2): 463–473. doi:10.1006/jeth.2001.2952.

Halsnæs, K., P. Shukla, D. Ahuja, G. Akumu, R. Beale, J. Edmonds, Christian Gollier, *et al*. 2007. 'Framing Issues'. In *Climate Change 2007 – Mitigation of Climate Change: Working Group III Contribution to the Fourth Assessment Report of the IPCC*. First edition. Cambridge University Press.

Holling, C. S. 1973. 'Resilience and Stability of Ecological Systems'. *Annual Review of Ecology and Systematics* 4: 1–23.

Holling, C. S. 2001. 'Understanding the Complexity of Economic, Ecological, and Social Systems'. *Ecosystems* 4(5): 390–405.

Jamieson, Dale. 1996. 'Ethics and Intentional Climate Change'. *Climatic Change* 33(3): 323–336. doi:10.1007/BF00142580.

Jonas, Matthias, Gregg Marland, Volker Krey, Fabian Wagner, and Zbigniew Nahorski. 2014. 'Uncertainty in an Emissions-Constrained World'. *Climatic Change* 124(3): 459–476. doi:10.1007/s10584–10014–1103–1106.

Knight, F. H. 1921. *Risk, Uncertainly and Profit*. Hart, Schaffner and Marx.
Kuhn, T. S. 1962. *The Structure of Scientific Revolutions* 1. University of Chicago Press.
Lenton, T. M., Hermann Held, Elmar Kriegler, Jim W. Hall, Wolfgang Lucht, Stefan Rahmstorf, and Hans Joachim Schellnhuber. 2008. 'Tipping Elements in the Earth's Climate System'. *Proceedings of the National Academy of Sciences* 105(6): 1786–1793. doi:10.1073/pnas.0705414105.
Lomborg, Bjørn. 2007. *Cool It: The Skeptical Environmentalist's Guide to Global Warming*. Cyan and Marshall Cavendish.
Lucas, Robert E., Jr. 1978. 'Asset Prices in an Exchange Economy'. *Econometrica* 46(6): 1429–1445. doi:10.2307/1913837.
Luke, T. W. 2010. 'Geoengineering as Global Climate Change Policy'. *Critical Policy Studies* 4(2): 111–126.
Mabey, N. 2011. *Degrees of Risk: Defining a Risk Management Framework for Climate Security*. Third Generation Environmentalism Limited.
MacCracken, M. C. 2009a. 'Beyond Mitigation: Potential Options for Counter-Balancing the Climatic and Environmental Consequences of the Rising Concentrations of Greenhouse Gases', *Policy Research Working Paper Series*, World Bank; available online at http://papers.ssrn.com/sol3/papers.cfm?abstract_id=1407956, accessed 23 September 2015.
MacCracken, M. C. 2009b. 'On the Possible Use of Geoengineering to Moderate Specific Climate Change Impacts'. *Environmental Research Letters* 4(4): 045107. doi:10.1088/1748–9326/4/4/045107.
Malone, Elizabeth, and Steve Rayner, eds. 1998. *Human Choice and Climate Change: What Have We Learned? V*. 4. Battelle Press.
Mason, Paul. 2011. 'Keynes vs Hayek – The LSE Debate'. Analysis, BBC Radio 4. www.bbc.co.uk/podcasts/series/analysis#playepisode3.
Moreno-Cruz, J. B, and David W. Keith. 2010. *Climate Policy under Uncertainty: A Case for Geo-Engineering*. http://works.bepress.com/cgi/viewcontent.cgi?article=1000&context=morenocruz&sei-redir=1#search=%22moreno+cruz+keith+uncertainty%22.
NAS. 2015. *Climate Intervention: Reflecting Sunlight to Cool Earth*. National Academy of Sciences. www.nap.edu/catalog/18988/climate-intervention-reflecting-sunlight-to-cool-earth.
NERC. 2010. '"Experiment Earth?" Public Have Their Say on Technologies to Reduce Global Warming'. September 9. www.nerc.ac.uk/press/releases/2010/35-experiment.asp.
Ott, K., G. Klepper, S. Lingner, A. Schäfer, J. Scheffran, D. Sprinz, and M. Schröder. 2004. 'Reasoning Goals of Climate Protection. Specification of Article 2 UNFCCC'. *Research Report* 202(41): 252.
Pryke, M. 2011. 'Introducing Andrew Haldane'. *Journal of Cultural Economy* 4(4): 365–369.
Ramsey, F. P. 1928. 'A Mathematical Theory of Saving'. *The Economic Journal* 38(152): 543–559. doi:10.2307/2224098.
Rittel, Horst W. J., and Melvin M. Webber. 1973. 'Dilemmas in a General Theory of Planning'. *Policy Sciences* 4(2): 155–169.
Sarewitz, Daniel, Roger A. Pielke, and Radford Byerly, eds. 2000. *Prediction: Science, Decision Making and the Future of Nature*. Island Press.
Schelling, Thomas C. 1995. 'Intergenerational Discounting'. *Energy Policy* 23(4–5): 395–401. doi:10.1016/0301–4215(95)90164–90163.

Shackle, G. L. S. 1953. 'The Logic of Surprise'. *Economica*. New Series, 20(78): 112–117.
Stern, N. H. 2007. *The Economics of Climate Change: The Stern Review*. Cambridge University Press.
Taleb, Nassim Nicholas. 2007. *The Black Swan: The Impact of the Highly Improbable*. Allen Lane.
Van Vuuren, D. P., J. Lowe, E. Stehfest, L. Gohar, A. F. Hof, C. Hope, R. Warren, M. Meinshausen, and G. K. Plattner. 2011. 'How Well Do Integrated Assessment Models Simulate Climate Change?'. *Climatic Change* 104(2): 255–285.
Walker, J., and Melinda Cooper. 2011. 'Genealogies of Resilience From Systems Ecology to the Political Economy of Crisis Adaptation'. *Security Dialogue* 42(2): 143–160.
Weitzman, Martin L. 1998. 'Why the Far-Distant Future Should Be Discounted at Its Lowest Possible Rate'. *Journal of Environmental Economics and Management* 36(3): 201–208.
Weitzman, Martin L. 2001. 'Gamma Discounting'. *American Economic Review*, 260–271.

6 Geoengineering – complexity in policymaking

> The domain of our ignorance is vastly greater than the domain of our knowledge, and if we implicitly or explicitly plan on the presumption of sufficient knowledge, we can be certain that failures will occur.
>
> (Holling 1977, 115)

The previous chapters have characterised the ecosphere as a complex adaptive system in which climate change is framed as a situation that demands adaptive management using techniques based on systems thinking, rather than as a problem that can be solved by reductionist thinking heavily reliant on predictive technologies. In this chapter I will examine the extent to which policymakers have adopted systems thinking as the framing for their deliberations generally and on climate change and geoengineering in particular. Anecdotally it would be surprising to find much evidence of this in policymaking because if it were common practice one would expect to see it reflected in media reports of government policy formulation, with comparisons of results from multiple pilot projects carried out in parallel, politicians explaining that they were devising policy heuristically rather than ideologically, and heralding policy failures as vital and welcome parts of the learning process whereby more apt policies emerge. This seeming absence is illustrated by an exchange in a UK Parliamentary environmental select committee report.[1] Recognising the importance of coming to terms with 'the domain of our ignorance' as referred to it in the epigraph above, the witness, Prof. Tom Horlick-Jones of Surrey University, explained the nature of heuristics as 'a way in which one uses quantified risk assessment [as] a guide to thinking more clearly about the issues' and that '[i]t gives you a guide to your own ignorance' even though it doesn't provide 'uncontaminated truth about the world'. Eric Pickles (a decade prior to being elevated to Secretary of State for Communities and Local Government) remarked that 'We are all fairly simple folk around this table' and promptly took the discussion off in another direction. It seems Mr Pickles wasn't interested in being guided around his own ignorance.

Instead, again anecdotally, we have a media that pounces on policy changes as 'U-turns' and signs of indecision and weakness, and politicians who seek to present themselves to their electorate as steadfast, masterful and unswerving, and who use every ruse available to deflect and diminish negative reaction to those policy changes that the harsh reality of evidence does periodically force upon them.[2] If this anecdotal view is in fact reasonably accurate, searching for examples of systems thinking in policymaking will be a challenging task. Nevertheless, if, as I argue, it is important that system thinking be adopted at the heart of policymaking in general and long term environmental policy in particular, it behoves me to show somewhat more formally that it is not already happening, or if it is, it is by no means a widespread approach. This chapter is limited to this narrow objective.

I have approached it in two ways. Firstly by means of a review of three documents that examine the use of systems thinking in the public sector. These documents are all from academic sources but are not peer reviewed for publication. The first is a paper published in 2011 by USAID (Ricigliano and Chigas 2011) and concerns the assessment of 'violent conflict for purposes of development planning and program design'; second is a report prepared by Cardiff University's Lean Expertise Research Centre (LERC) in 2010 to evaluate systems thinking in the UK public sector on behalf of the Wales Audit Office; and third is a paper by Lebcir (2006) that examines the history of systems thinking in the public health care sector. None of these papers is directly concerned with environmental management but they do provide useful overviews of the take up of systems thinking in the public sector. The cases studied in these papers are far from the only applications of systems thinking in the public sector, for example, several others are referred to by Lempert *et al.* (2003) and Walker *et al.* (2013). However, they suffice to demonstrate the tentative but committed experimentation with adaptive management.

The second approach is a critical review of five recent documents, four of which specifically address geoengineering while the fifth deals more generally with climate change. None of the four geoengineering documents is a government policy document *per se* (because to date no government has an established policy on geoengineering that amounts to anything more than 'a watching brief', NERC 2010, 15[3]). Nevertheless, they are all written with a view to influencing policymakers and have a provenance that makes them credible proxies for the framing by policymakers of climate change and the potential role of geoengineering as a response to it. The four geoengineering reports are the two reports from the UK/US joint House of Commons and House of Representatives enquiry into geoengineering that reported in 2010; the report issued in 2012 following the 2011 IPCC expert meeting on geoengineering; and finally the 2010 US Bipartisan Policy Committee report. The fifth document, the Stern Review, was commissioned by the UK government specifically to advise policymakers about the long term economic dimension of climate change and whether a policy of emissions abatement was economically justified.

The chapter concludes with a brief summary of the evidence for systems thinking in these documents and, given that prediction will be seen to play an important role, I also assess them according to Sarewitz *et al.*'s criteria for the appropriateness of their reliance on prediction: timeliness, necessity, accessibility, compatibility and distraction.

Case studies of systems thinking in public policy

USAID

The analysis in this report from the US government international aid agency is based on over sixty conflict assessments since 2002 and directly recognises the challenges of complexity (Ricigliano and Chigas 2011, 2). They understand systems thinking as 'a way of understanding reality that emphasizes the relationships among a system's parts rather than simply the parts themselves'. They distinguish between simple, complicated and complex systems according to a combination of the degree and ease of predictability. For both simple and complicated systems there are high levels of predictability although establishing causal chains is more challenging for the latter. In complex systems there are high levels of uncertainty and low levels of predictability. They highlight connectedness, non-linearity, feedback, patterns, and emergence as being the primary attributes of both complicated and complex systems. At the complex end of this spectrum '[a]ttempts to define and control outcomes often result in failure, because in complex contexts solutions cannot be imposed, but arise from the circumstances' (Ricigliano and Chigas 2011, 3).

Ricigliano and Chigas also stress that systems thinking is an adjunct and not an alternative to analytical thinking. In particular they list several shortcomings of the conventional approach to conflict reduction that are also redolent of its limitations in respect to climate change. Their list includes fragmented problem definition and failure to integrate results, biased, narrowly focused and partial analysis designed to serve vested interests, and failure to prioritise. Under three headings, *Comprehensiveness* (seeing the system as a whole yet keeping it simple), *Comprehensibility* (focussing on key dynamics and avoiding long to-do lists), *Portability* (creating a continuously reflexive learning environment), they explain how to apply systems thinking to conflict situations. They identify systems mapping as a key tool in understanding a system's dynamics and boundaries (the obesity diagram in Chapter 4 is an example of such a map).

Much of the paper is concerned with the mechanics of producing system maps and they use several case studies as illustrations (e.g. ethnic tensions in Kosovo, Sri Lanka civil war, Kiribati's failed attempt to address serious degradation of local fish stocks). In their conclusion they comment that:

> Systems thinking is a way to produce rich assessments (narratives) of complex environments that facilitate effective program planning, implementation, monitoring and evaluation, and learning from experience.

> Finally, it is appropriate to remember that one of the hallmarks of systems thinking is emergence – or the tendency for new strategies and outcomes to arise that were not necessarily contemplated at the start of a process.
>
> (Ricigliano and Chigas 2011, 28–29)

Albeit that USAID's engagement with systems thinking may be incipient, it is indicative of a recognition in public policy circles that the conventional predict and control methodology is ill-equipped to cope where the challenges are 'difficult, recurrent or intractable [and] whose solution is not obvious and [...] involve[s] complex issues and a need for multiple actors to coordinate and see the "big picture," not just their part in it' (Ricigliano and Chigas 2011, 3). All these conditions apply to responding to global climate change.

LERC

Cardiff University's LERC study was based on three case studies in UK local government; one involving the delivery of facilities for the disabled and the other two, the administration of housing benefits. The details of the case studies need not detain us because their minutiae are far removed from those of managing global climate change. However, LERC's conclusions from the study are informative and potentially of wider application.

They grouped the themes emerging from the application of systems thinking into four categories: a) providing a framework for change; b) impact of targets; c) wider system implications; and d) sustainability of changes. Their first conclusion was that the process of engaging with systems thinking refocused the participants on effectiveness in contrast to the prior emphasis on efficiency. This was enhanced by a recognition that ineffectiveness was as much a function of system design as it was of inadequate performance. This realisation encouraged the participants continuously to focus on iteratively designing out the propensity to fail, a process referred to as adaptive management. This process accepts that some degree of failure, or sub-optimality, is inevitable and seeks to ensure not only that the system is resilient to those failures, but also, and equally as importantly, that they inform further changes in system design to increase effectiveness.

In all three case studies there was clear evidence that the performance targets that were a feature of the earlier systems were driving perverse behaviours because 'meeting the target' became the primary system objective rather than delivering an effective service. Systems thinking reoriented normative behaviours in line with the real purpose of the system. A side effect of this was also significantly to improve efficiency, an important outcome now delivered as an epiphenomenon.

The systems thinking process also made participants aware of the impact their improvements had on other, ostensibly unrelated systems. For example, improvements in the delivery of facilities to the disabled reduced health and residential care costs, and those in housing benefits reduced demand for services at the Citizens' Advice Bureau.

The fourth theme, sustainability of the changes, was difficult to assess because in all cases the projects had yet to stand the full test of time. Nevertheless, the initial indications were that the active engagement of the staff at all levels, and their taking of responsibility for delivering effective services, were instrumental in changing their approach to their work, clients and colleagues, thus making the changes more sustainable. LERC likened this effect to that of Jidoka, a feature of Japanese lean management techniques where the engagement of workers raises their self-esteem by making the work experience less mechanical and dehumanised.

LERC are careful to warn that the conclusions drawn from these three case studies should not be extrapolated without further research. This is doubly so if we are to apply these experiences to the challenges of dealing with global climate change and the governance of geoengineering. Nevertheless, they demonstrate that systems thinking proved to be a valuable means of transcending tribal behaviours within interrelated public service departments.

Lebcir

Dr Lebcir's study is also drawn from the tradition of management practice and much of his treatment repeats material discussed elsewhere. However, Lebcir also discusses the distinction between 'detail complexity' and 'dynamic complexity' and argues that much poor management derives from an excessive focus on the former and inattention to the latter. He describes *detail complexity* as selecting a choice from a number of static options. This may be difficult because of both their number and individual complexity, although this choice can be aided by modelling. *Dynamic complexity*, on the other hand, arises when:

> (a) the short and long term consequences of the same action are dramatically different; (b) the consequence of an action in one part of the system is completely different from its consequences on another part of the system; and (c) obvious well-intentioned actions lead to non-obvious counter-intuitive results.
>
> (Lebcir 2006, 7)

Understanding dynamic complexity is important, he continues, because it allows 'the identification of leverage points in a system to improve performance and avoid policy resistance'.

Dynamic complexity is driven by feedback loops, time delays between action and effect, and non-linear relationships between system elements. He describes *unintended consequences* as little more than effects that decision makers failed to predict 'as a result of flawed and incomplete conceptualisation of the feedback loops' (2006, 8). He explains the distinction between positive or reinforcing feedback loops, and negative or balancing feedback loops and regards all actions 'merely as influences trying to shift the balance of power among the system's feedback loops' (2006, 8).

The time delay between action and effect, particularly where it is attenuated, is commonly a source of over-stimulation of the system as more action is taken in the absence of evidence that earlier actions have produced the desired effects. The accumulation of latent effects eventually results in an excessive response that is greeted by an equally excessive reaction. This results in oscillating behaviour as the system is sequentially over- and under-perturbed.

Lebcir argues that non-linearity is the primary source of unpredictable system behaviours particularly where the non-linearity arises from path dependence, or threshold effects where small inputs can result in phase changes in a system's behaviour. Furthermore, he explains that in their decision making, humans display a *bounded rationality* with two dimensions. The first derives from the limited information processing capabilities of the human mind that oblige us, when faced with the complexity of the real world, to focus on a reduced amount of information and simplify our mental cause-effect maps by using linear thinking and ignoring the side effects of decisions. This introduces structural uncertainty into our mental models.

The second bound of rationality is due to the cognitive skills and memory limitations of the human mind that render us incapable of computing all the consequences of our actions even were we to have perfect information about the cause-effect maps of a feedback system (2006, 10). Lebcir notes that the empirical evidence suggest that the effects of bounded rationality are particularly severe when humans are confronted by dynamic complexity and that we routinely 'ignore feedback structures, do not appreciate time delays between actions and consequences, and are insensitive to the non-linearities between a system's elements as the system evolves over time' (2006, 10).

Lebcir's distinction between detail and dynamic complexity can be applied with considerable effect to climate change and geoengineering. In particular his observations about sequential over- and under-perturbation of the system, and the effects of bounded rationality provide valuable insights into understanding the dynamics of climate policy formulation.

Discussion

The USAID example aligns more closely with global climate change in that it operates at an international scale, albeit not global in the sense of 'everywhere at once' as would be the case with geoengineering, and it deals with extreme complexity deriving from the nature of the problems with which USAID is involved. The LERC and Lebcir papers draw more heavily on the branch of systems theory that comes through business and enterprise management (e.g. Checkland, Deming, Ackoff, Seddon) than on that from environmental management (e.g. Holling, Folke, Gunderson). A further key difference between climate change and all these examples is their temporal extension. Few wars extend beyond a few years, perhaps in the extreme, a few decades. The temporal scale of the delivery of housing benefits and disabled facilities is typically measured in days and months. Climate change, on the other hand, unfolds over long time scales. Policies implemented today may not produce measurable results for many decades or even centuries.

Despite their limitations, the examples are of interest because they all draw on the same theoretical foundation of CAS theory. In all these applications, the central feature is the focus on the complex dynamics of the relationships between the system elements and their emerging properties, rather than on the ontology of the elements themselves. Although they might not refer to wickedness, reflexivity or paradigm shifts, they embrace these concepts with terms such as complexity, feedback loops, non-linearity, bounded rationality, and Jidoka.

The growing interest in systems thinking in both academic and commercial contexts has resulted in the UK government being advised in many different policy arenas of its benefits in overcoming the limitations of more linear processes when attempting to build protection from and resilience to unknowable future threats. Two recent examples are evidence from Voudouris to the 2011 Energy and Climate Change committee report on the UK's energy supply;[4] and a 2012 submission to the Defence Select Committee on cyber-security from the Institute for Security & Resilience Studies at UCL.[5] Notwithstanding these recommendations, the adoption of systems thinking in key policy areas has yet to happen. Certainly in relation to geoengineering and climate change, this conclusion is largely confirmed in the review of the five policy papers discussed in the next section.

Geoengineering policy advice

House of Commons Report (2010)

This Committee's brief was to consider the governance issues around geoengineering. The Committee members were all elected UK Members of Parliament, several of whom had doctorates in a range of natural and social science subjects but none with any academic experience related to climate change. It heard evidence from a number of participants including an international group of scientists comprising several of those who had authored peer reviewed papers on geoengineering, and some of whom participated, either as witnesses or authors, in all four geoengineering reports considered here. In addition, there were UK politicians and ETC, a Canadian environmental advocacy group whose report entitled *Geopiracy* is strongly opposed to all forms of geoengineering (Marshy 2010). This Committee was more concerned with the regulation of geoengineering than with research, which was the remit of the co-ordinated parallel process in the US House of Representatives' committee. Accordingly, this analysis focuses on the framing of the science and its geopolitical implications rather than on a detailed analysis of the geoengineering technologies and their wider socio-political implications.

After two sections of general background material, the third part of their report discusses the nature of the regulatory challenges presented by geoengineering. Neither the evidence presented to them nor their discussion of it addressed systemic issues but focused on the narrower question as to whether, as a matter of principle, geoengineering research warranted governance at national or supra-national level and to what extent it was already regulated by existing treaties. It concludes that there is a need to establish an international regulatory process for geoengineering.

The fourth part of the report engages with what form that governance might take and how it might be instituted. They endorse the five Oxford Principles[6] but make only the most general statements about how they might be pursued. They argue for international geoengineering regulation to be managed through the auspices of the UN. They reject the application of the 'precautionary principle' arguing that it 'should never be considered a substitute for thorough risk analysis which is always required when the science is uncertain and the risks are serious' (House of Commons 2010, para. 86). However, there is no discussion in the report about how to operationalise risk analysis 'when the science is uncertain'.

Their conclusions in this section recommend that research into geoengineering be allowed to proceed subject to a prior formal governance regime being established provided it meets three criteria. Firstly that it be conducted in accordance with an internationally agreed set of principles such as the Oxford Principles, secondly that it has negligible or predictable environmental impact, and thirdly that it has no transboundary effects. There are, however, some difficulties with these recommendations that the application of systems thinking might have exposed. Establishing a set of preconditions for the international governance regime may create unnecessary obstacles if others have different views. For example, there is almost inexhaustible scope for both procrastination and prevarication in seeking international agreement to the meanings of 'negligible' and 'predictable' according to the potential vulnerability of the actor and the differences in their understanding of and confidence in the underlying science. Similarly a requirement proscribing all transboundary effects seems unnecessarily absolute and pre-emptive. According to Beck, one of the inevitable consequences of globalisation in its myriad manifestations is the extent to which activities undertaken wholly within the bounds of one nation-state can influence those elsewhere, sometimes positively sometimes not. The transboundary implications of geoengineering are specifically highlighted by Humphreys (2011). An environmental policy that disallows research that might have transboundary effects is likely to be one that, in effect, limits research to the ineffective.

The Committee does highlight Sir David King's observation that explicitly identifies the extraordinary complexity deriving from 'emergent properties that are not always easy to predict' (House of Commons 2010, para. 92). But they do not develop this with any further consideration of how this unpredictability might affect the challenges of governance other than to recommend that 'regulatory measures [be] imbued with a high level of flexibility to be able, for example, to encompass new technologies as they emerge' (House of Commons 2010, para. 103). However, this focus on emergent technologies ignores the possibly greater imperative of an openness to recognise and encompass new climate induced threats to the ecosphere as they emerge; this is the territory of uncertainty and surprise. This is particularly so where, in keeping with Beck's notion of reflexive modernisation, those threats are emergent properties of the technologies employed to remedy earlier threats, technologies whose continuation is then promoted and protected by a burgeoning set of powerful vested interests.

While this report does not pretend to engage in the detailed design of a regulatory regime for geoengineering, its framing of the issue perpetuates the normal science paradigm (Kuhn 1962) of working within the boundaries of what is known and what is known not to be known. Although there is some recognition of the challenges provoked by complexity, the report's recommendations are devoid of any suggestions that might embrace uncertainty over the spatial and temporal scales relevant to any timely and effective response to climate change.

House of Representatives Report (Gordon 2011)

This report focused on the research implications of geoengineering. Again, the witnesses interviewed by the panel were largely drawn from the same small group of US and UK academics active in the area of geoengineering. The main body of the report is a catalogue of existing US research facilities engaged in activities with some relevance to geoengineering.[7] Many of the programmes described include as key objectives the prediction of future climate states and other environmental variables including social impacts and biodiversity. For example, the National Ecological Observatory Network undertakes research that will enable it to 'quantify forces regulating the biosphere and predict its response to change' (Gordon 2011, 10). The National Oceanic and Atmospheric Administration 'conducts broad ranging research into complex climate systems with the aim of improving our ability to understand these systems and predicting climate variation and change over a range of temporal and spatial scales' (Gordon 2011, 10). The US Global Change Research Program has as one of its stated aims 'to predict and reduce uncertainty in projections for climate change in the future' (Gordon 2011, 34). The issue for this Committee was how best to marshall the considerable resources available in the US already engaged in research to enhance the predictability of climate variables. There is no evidence of any argument that a reliance on prediction might be misplaced, or that research should be undertaken to examine how complexity theory might be considered, whether as an alternative or complement to the predict and control framing of climate change and geoengineering.

Two specific recommendations are directed at downgrading the research priority of space and desert based solar reflectors and another is directed at raising the research priority of localised SRM. But neither the submissions on which these recommendations were based, nor the deliberations around them placed any priority on the systemic implications of these geoengineering options. Their rejection and promotion, respectively, turn on a narrow set of normal science first order considerations. This failure to account for the wider system implications of climate interventions is an exemplar of the action of a *global controller*[8] who, by imposing his will, limits possibilities for the creation of potentially more sustainable ways of managing the climate situation.

The overall tenor of the recommendations in this report is very general and it is quite possible that others responding to them in due course could do so in ways that do embrace systems thinking. Nevertheless, the framing that places the role of prediction at the heart of the research process is clearly shown by their call for 'exhaustive efforts [to] be made to identify and avoid the most dangerous [unintended consequences] before a climate engineering program is tested or deployed at any scale' and that the risks of large scale field testing can be eliminated by appropriate computer climate modelling (Gordon 2011, 20). The implication here is that a low risk geoengineering programme could be established in virtual digital space and then deployed fully formed in the real environment. A moment's reflection on the central lessons of CAS theory, and in particular the nature and role of emergent properties, suggest that this is a fanciful aspiration. Even were we able to identify all the dangerous unintended consequences, rendering the future devoid of surprises, ranking them to establish the 'most' dangerous that would need to be avoided would introduce both computational and normative questions open to much contestation.

The framing here is that responses to climate change need only address the contingencies (whether intended or not) that can be conceived prior to their occurrence, and this is best achieved by improving predictions so as to eliminate surprises. By contrast the CAS approach is to accept that there will be surprises and the best way to cope with them is to build system resilience that mitigates the risks from catastrophic events by both reducing their likelihood and improving our capacity to absorb them even though we may not know in advance how, when or where they will arise. From this report it seems that the House of Representatives believes that we can eliminate surprises by improved prediction.

IPCC expert committee on geoengineering

The IPCC is an international collaboration of scientists not a policymaking entity. Nevertheless, because of its status as a UN agency respected by policy-makers as an authoritative source about all aspects of climate change, it acts to some extent as an indicator for policymakers' framing of climate change.

The IPCC expert committee on geoengineering (IPCC 2012) was tasked with scoping the topic of geoengineering for its inclusion in the then forthcoming IPCC AR5 report in such a way as to promote some uniformity in the language and treatment of the subject by the many independent scientists who were likely to contribute papers. As such it was primarily a discussion about definitions and the classifications of the many geoengineering methods, their mode of operation, the nature of their effects and side effects, and the criteria by which they might be compared. This review focuses only on the four pages of the Summary of the Synthesis Report that highlights those issues the Committee felt worthy of extracting from the many posters and papers submitted to the session and making up the bulk of the remaining 104 pages. The narrow question with which I am concerned here is the extent to which their report recommends a CAS approach to geoengineering to authors contributing papers on geoengineering for AR5.

There are two specific references in this brief report that tilt somewhat tentatively towards CAS theory. Firstly they note that geoengineering should not be considered in isolation from climate change itself and other responses to it. This is a significant, although limited move away from framing geoengineering technologies as solutions to a problem, and towards them being part of managing a bigger system. The second is an explicit suggestion that there might be advantages in 'a holistic comparison of response options' and a framework is proposed whereby this might be undertaken not only for geoengineering but also for emissions abatement and adaptation.[9] However, given that the predict and control approach is the *de facto* default paradigm (for example, as evidenced in the many submissions to this Committee (e.g. Buesseler *et al.*; Peter; and Smith, all in IPCC 2012)), one would expect a shift to a systems thinking approach to be evidenced by some commentary about the need for making such a shift. Despite the references to holism there is no such discussion. Only MacCracken hints at a different way forward when he notes that:

> the more we have learned about the climate, the more we have come to understand the hubris involved in contemplating upsetting its many valuable and beneficial intercouplings.
>
> (MacCracken in IPCC 2012, 55)

However, MacCracken is not recommending a moratorium or banning of geoengineering, rather he proposes a cautious heuristic move to enhance our understanding without recklessly causing a climate 'upset'. This is to be done by adopting an explicitly adaptive management approach. He argues that 'there is a much greater likelihood that climate engineering can play an important positive role if started up on a limited scale than if there is an immediate jump to the global scale'. He outlines five local scale projects to address specific climate change issues.[10] While not expressing himself in explicitly systems thinking terms, his proposals move in that direction.

The reference to a holistic approach suggests some systems awareness but the report's parenthetic definition of *holistic* is extremely brief – 'e.g. inclusive of all potentially relevant aspects' (IPCC 2012, 4). But systems thinking does not proceed from a concern to consider 'all potentially relevant aspects'. As Cilliers explains:

> We cannot consider life, the universe, and everything in its totality all the time. We need limits in order to say something.
>
> (Cilliers 2002, 81)

Deciding what is 'potentially relevant' is likely to be contentious, and increasingly so the more extensive the spatial and temporal frame of reference, and may undermine consideration of the substantive issues. It may be that the authors of this section of the report intended that 'relevance' be understood in a more nuanced manner than merely 'bearing on, or connected to'[11] but the paucity of discussion leaves the matter open. In the absence of any other reference to CAS theory – heterogeneity, emergence, autonomous processes of selection and enhancement, and resilience – it would be an unjustified reading of this report to regard it as engaging seriously with a systems thinking approach for the IPCC's assessment of geoengineering. However, there is clear, albeit inchoate evidence that an awareness of the benefits from a systemic view is emerging.

IPCC AR5 was published in 2014 and all three working group reports deal extensively with geoengineering. For the most part the treatments are in the form of comprehensive literature reviews, highlighting the extent of the contestation and the knowledge gaps that pervade the geoengineering discourse. They make no concrete policy recommendations and while the predict and control framing is dominant with many references to geoengineering as an emergency response to extreme climate change, there are also some references to systemic considerations. It is also noteworthy that only in one of the three summaries for policymakers is there any mention of geoengineering, and then dismissively. This indicates a determination to suppress debate about geoengineering by the policymakers responsible for summarising the scientists' full report text for public consumption. Notwithstanding MacCracken's views expressed to the earlier Expert Committee, there was no evidence in IPCC AR5 of a shift towards an heuristic and systemic approach to geoengineering research.

Bipartisan Policy Center Report of a Task Force on Climate Remediation Research (2011)

The BPC was founded in 2007 by four high-ranking US Senators, two from each party, as a bipartisan think tank.[12] Its reports are not those of policymakers but, as with the IPCC, it has a certain pedigree that suggests that it understands how to present an argument in such a way as to appeal to the political establishment and in particular to the US Administration. The BPC report presents a number of proposals concerning research into geoengineering. The eighteen panellists brought together for this report include several scientists eminent in the field of geoengineering, individuals who have also contributed to the IPCC expert committee and to the Royal Society's SRM governance project and its *Geoengineering the Climate* report (Shepherd *et al.* 2009) including Caldeira, Keith, and Shepherd. In addition there were a number of eminent academics from related disciplines and former diplomats and US government officials.

The purpose of the report is to advise the US government how to improve 'its understanding of climate remediation options[13] and how it should work with other countries to foster procedures for research based on that understanding' (BPC 2011, 3). They propose six principles that should govern geoengineering research; they overlap with the Oxford Principles: a) public good; b) moratorium on field tests; c) broad expert and civic engagement; d) transparency; e) international co-ordination; and f) adaptive management. They state:

> The environmental, scientific, technological, and social context for climate remediation research is likely to evolve significantly over time in unpredictable ways. Federal research programs should be required to review those changing conditions on a regular basis. The program must establish a transparent process for changing focus, direction, or research procedures in response to changing conditions. Institutions involved in climate remediation research should have the responsibility to evaluate assumptions and to test predictions against new information and actual observations.
> (BPC 2011, 14)

There is no discussion about how this might be done but it is an open invitation to those engaged in this research to recognise the importance of some form of systems thinking. This is a significant recognition that the management process itself must be adaptive, quite apart from any actions that emerge also enhancing adaptivity in the environment.

When discussing the institutional structure necessary to promote geoengineering research in the US, the report notes the diversity of agencies already involved and observes that while this presents certain organisational challenges it is, if correctly managed, a potential asset. They suggest that the co-ordination role should fall within the White House because it is the only entity that has a purview of the entire research activity and the ability to intervene both nationally and internationally. This introduces an interesting tension according to whether the White House operates in a co-ordinating role, or alternatively, more in keeping with US geopolitical hegemonic aspirations, as a literal global controller leading the international effort. In this latter case, the effect could be to suppress, albeit unintentionally, the diversity necessary to promote innovation. Both modes are represented in the report.

The significance of a learning culture at the heart of CAS theory, whereby new knowledges from multiple sources can be combined into novel ideas, is reflected in the following observation by the BPC:

> Establishing a coherent research program requires a coordinated effort to draw effectively on each agency's strengths and to ensure those strengths are applied in an integrated fashion across a range of research activities, including natural and social sciences, engineering, and the humanities.
>
> (BPC, 2011, 18)

However, in the section prioritising topics for the research agenda, the predict and control paradigm prevails unchallenged. They present five knowledge gaps worthy of further research, each of which concerns improving our understanding of different aspects of the physical climate system in order that more accurate predictions can be made so as to 'steer the science toward the answers'. Notwithstanding the earlier comments concerning the importance of including the 'social sciences, engineering, and the humanities' alongside the natural sciences, the BPC's research priorities relegate them to the margins. Furthermore, the desire to 'steer the science toward the answers' is quintessentially located in the predict and control paradigm and assumes the existence of 'answers' capable, at least in principle, of resolving the climate problem. This is a proposition entirely at odds with the central tenets of CAS theory.

The BPC report may be interpreted as being on the cusp of a change to a more systemic approach to addressing the challenges of climate change. On the one hand it recognises the importance of being transparent, capacious and adaptive in the face of changing circumstances and the need to build those features into the institutional structures supporting the research programme. Yet on the other, the priorities highlighted are firmly rooted in the perceived power of the natural sciences to produce 'answers' to intractable problems. This fails to register the complexity, the wickedness, the reflexivity that produce a constantly morphing set of challenges characterised by the inevitability of surprise for which no amount of prediction will prepare us. It remains to be seen whether the benefits of combining a systemic view with the conventional reductionist methodologies will prevail against the hegemonic attraction of predict and control.

Stern Review

The Stern Review is a document of some 650 pages. In this brief section, I limit myself to an examination of the extent to which it recognises the wickedness and complexity of climate change and accommodates them in its principal conclusion that 'the benefits of strong and early action on climate change outweigh the costs' (Stern 2007, i). Although the Stern Review no longer has quite the prominence that accompanied its original publication,[14] it provides an insight into the current application of economics to climate change matters. Stern makes no reference to geoengineering but is focused on the economics of emissions abatement and to a lesser extent of adaptation. This is unsurprising as the bulk of the work on this extensive paper was done just before the publication of Crutzen's 2006 editorial that brought geoengineering onto the policy agenda.

Stern addresses systems complexity in his review of the potential for adaptation to ameliorate the effects of climate change (Stern 2007, pt. V). He distinguishes between *policy-driven* and *autonomous* adaptation. Autonomous adaptation fulfils all the criteria for complex adaptivity. He explains that this is a local response by many individuals and organisations to local circumstances. Although he does not make the point explicitly, his text suggests that he is aware that emergence is also a feature of this adaptive behaviour. Policy-driven adaptation, on the other hand, is 'a deliberate policy decision', usually taken by public authorities rather than the private sector; policy-driven adaptation is the act of an exogenous actor – a global controller (although he does not use the term) (Stern 2007, sec. 18.2).

An issue not considered by Stern is whether an economic justification for responding to climate change is necessary. A parallel can be drawn with defence policy where vast amounts are spent in an attempt to secure national security. While individual expenditures within the defence budget may be assessed on a value-for-money basis, the overall defence budget is politically determined in a process that makes no attempt at a discounted cost benefit analysis that measures, for example, the expenditures on nuclear weapons against value of the perceived benefits they are supposed to provide. Epistemologically, as there is no control environment in which there are no nuclear weapons, there can be no evidence that not having nuclear weapons would make the world a less safe place. Policymakers do not argue that the existence of a large and well-equipped military may be as likely to precipitate conflicts as to resolve them, nor that the money spent on armaments might produce more welfare if spent elsewhere, or not spent at all. Yet, for climate change, policymakers demand economic justification despite the existential threat it poses being every bit as real as that from external security breaches.

Policymaking decentralisation

It is because Stern does not consider global scale emissions abatement to be feasible other than through centralised policy-driven decisions that he does not treat it as being subject to autonomous local variations. However, complex systems theory does not limit the size of such systems. The two hundred or so nation-states operating anarchically comprise a complex adaptive system. From that system many regional and local innovations would emerge, of which some would succeed and others fail, all combining to leave their traces on the continuously changing face of the ecosphere.

Moreover, the assumption that emissions abatement can only be delivered through a unified public policy implies the creation of a truly global global controller. This conclusion arises because if a centrally agreed emissions abatement policy were implemented locally or regionally, it would be subject to local variations from both natural and social factors. These variations would, through their multiple local interactions, form a complex of complex adaptive systems with their own emergent properties and therefore rapidly cease to constitute a centralised policy. It follows that a centralised policy must be globally uniform in order to suppress the emergence that would otherwise cause diversity. Furthermore, the suppression of emergence can only be achieved by the fiat of a global controller. However, at global scale it is unimaginable that any global controller could effectively suppress all emergent properties other than for a very short time, and possibly not even then. This suggests that Stern's premise that emissions abatement can only be delivered by centralised policy-driven adaptation is false. Indeed, it suggests the converse, that effective emissions abatement can only be delivered by some form of decentralised adaptive management process.

Efficiency and resilience

A second key systems issue is the relationship between efficiency and resilience. Throughout the Stern Review there are many appeals to efficient outcomes (there are some 600 mentions of *efficiency* and fewer than 100 to *effectiveness*). Efficiency, a concept central to economic analysis, implies minimum economic value of means for maximum economic value of ends. This is not a concept that fits well with CAS theory because the process of emergence, the source of all innovation in complex systems, is driven more by serendipitous juxtapositions than appeals to optimal use of resources.

Viewing economic cycles through the lens of adaptive cycles, a long bull market corresponds to the r-phase of growth and accumulation; the transition from the bull to the bear market describes the peaking of the K-phase quickly followed by the collapse, or *correction* as it is euphemistically known, of the Ω-phase. The recovery in the α-phase completes the cycle. CAS theory would suggest that in order to attenuate the r-phase, to defer or even avert a collapse, it is necessary to ensure resilience within the system. Resilience is enhanced by the redundancy that provides alternative pathways in the event of partial system failures; but redundancy is contra-indicated by efficiency. The word *redundant* implies waste and excess provision both of which are inimical to efficiency. In systems in which redundancy is designed-in expressly to improve system resilience, it is commonly referred to as an *overhead*, recognising it as an incremental cost. Many systems in nature are far from efficient.

For example, photosynthesis, that supports all aerobic life on the planet, has an efficiency of only 5 per cent.[15] Fail-safe systems (e.g. the Internet, aircraft, nuclear reactors) typically have built-in redundancy to ensure resilience so that in the event of a partial system failure, duplicate functionality is immediately available to avert total system collapse. But, as experienced with the Fukushima tsunami, the resilience of planned systems is only as good as the predictions for potential failure. Surprise challenges resilience. The pursuit of efficiency will always tend to undermine resilience and conversely, the pursuit of resilience will tend to reduce efficiency. However, as is well illustrated in the case studies examined in the LERC report, the tendency to reduce efficiency by increased structural redundancy can be more than overcome by operational factors that reduce operating errors and a reduction of demand for other services downstream, provoked by improved effectiveness upstream. It can be argued that long periods of growth and accumulation foster a degree of complacency towards the need for bearing the additional cost of redundancy, that becomes increasingly sacrificed as the growth phase progresses. Once there is insufficient flexibility in the system to absorb an unexpected event, the entire system becomes vulnerable to destabilisation. The 2008 banking crisis and the BP Deepwater Horizon oil disaster can both be understood, in part, as failures to invest in redundancy driven by an over-zealous drive for economic efficiency.

This tension is at the heart of predict and control. If it is believed that an outcome can be predicted sufficiently reliably to control it, resilience need only cater for those risks that are predicted and for known uncertainties. Planning for additional resilience to cater for surprises is tantamount to acknowledging that the prediction is unreliable. When considering the social impacts of today's policy options unfolding into the distant future, economists have yet to find a way of resolving this tension other than the controversial practice of discounting. It follows that the quest for economic efficiency is unlikely to be a prudent framing for robust responses to climate change. This is even more so because, as Stern himself recognises, there are many important human values not captured by economic variables and therefore not amenable to discounting.

Summary of policymakers' documents

From these five sources, the conventional predict and control approach is demonstrably dominant. That systems thinking might be a more apt way of framing climate change and geoengineering is only indicated by MacCracken's suggestion that empirical geoengineering research should start early and small and be scaled up as we learn more, allowing us to contain, even reduce, the associated risks, and the BPC's recommendation for adaptive management. However, this inchoate recognition is overwhelmed by a more general determination to prioritise further research in the natural sciences in order to improve predictability by increasing knowledge of the climate system, and thereby reduce risk – the hallmarks of a predict and control methodology. It seems that the anecdotal evidence that systems thinking is rare in public policy has some empirical support in the geoengineering discourse.

To predict or not to predict?

The evidence from USAID, LERC and Lebcir suggests, albeit on a small scale to date and far from conclusively, that systems thinking can be applied to public policy with positive results. However, if predict and control is dominant, is there evidence that it is an inappropriate methodology? In this section I will apply Sarewitz *et al.*'s criteria[16] for assessing the value of prediction to support policy formulation, to the problematic of climate change and geoengineering.

Are timescales for prediction production and political action commensurate? The fundamental problem for climate change prediction producers is that the timescales relevant to their predictions run at their shortest over decades and at their longest over centuries and even millennia. If action depends upon validating predictions with empirical evidence there is an irresolvable problem if the evidence will not be available until long after the policy needs to be implemented. This impasse illustrates the point made earlier that using explanatory predictions to test an hypothesis is fundamentally different from using decision-making predictions. These temporal dynamics suggest that prediction of future climate states fails to meet the first of Sarewitz *et al.*'s criteria.

Are accurate predictions necessary for decision makers or do they have viable alternatives? This criterion could be replaced by the simple question: if policymakers had no predictions at their disposal, would they be unable to devise policy likely to improve outcomes? The argument pursued throughout this book is that an heuristic approach involving multiple incremental steps from which empirical evidence is used gradually to adapt to the changing circumstances so as always to improve resilience as the primary protection against surprises, is a route that is always open to policymakers. It is a policy path quite different from the customary target driven policy processes as exemplified by the UNFCCC's 2°C temperature increase limit, but one, as illustrated by the examples discussed earlier in this chapter, that has potential to deliver continuing results even as the circumstances change. Adopting Wigley's 'buying time' framing for geoengineering (2006) provides precisely the basis needed to develop policy heuristically without having to rely on predictions and without having to incur undue risk. Climate change and geoengineering predictions fail Sarewitz *et al.*'s second criterion because policymakers do have viable alternatives to prediction-based policies.

Can the predictions be readily interpreted by non-experts to avoid their misuse or abuse? There is little in the five documents that addresses this point directly. Apart from some recognition that wider publics need to be engaged in the policy formulation process, there has been little discussion of the practical issues this might entail or the role that prediction may play. However, there is increasing evidence of highly qualified scientists working for the media and advocacy bodies (e.g. Greenpeace, Friends of the Earth, WWF) and their role in the climate change and geoengineering debate is to act as a counterweight against the hegemony of the academic climate scientists and ideologically driven politicians. Although geoengineering is still no more than a concept, there is ample evidence of these organisations fulfilling this role and continuing to work to provide an interface between the research, the policies and the public. There seems little reason to suppose that informed and responsible independent intermediaries could not ensure that predictions were not abused and misused. It might be argued that there is a greater risk that vested interests might abuse and misuse the predictions, but this can hardly be regarded as a failure of the predictions themselves. I conclude that climate change and geoengineering predictions are no more susceptible to abuse and misuse than any other element of the climate change debate.

Are the criteria for success in scientific prediction sufficiently similar to those that apply in policymaking to avoid confusion and controversy? This criterion is difficult to assess because international policymaking has to date been largely ineffective, as is demonstrated by the unabated increase in CO_2 emissions. It is therefore not yet possible to know what the criteria for success in policymaking are. Moreover, although the considerable increase in computing power in recent years has undoubtedly resulted in more precise predictions, it will not be possible to know how accurate they are until the distant future they predict arrives (or doesn't). Judgement on this criterion must remain open until suitable data become available, by which time, of course, the moment for action will have long passed.

Is there a risk that a focus on prediction will drain financial and other resources away from other activities that might better address the issue at hand? It is very clear from the documents reviewed that investment in improving predictive capacity is seen by most as a high priority. Insofar as prediction is an inappropriate tool on which to rely for the purpose of climate change and geoengineering policy formulation, it must be the case that that incremental investment could be better directed. Moreover, because other research activities are deferred in expectation of more accurate predictions, it must also follow that resources for that other research are indirectly drained. It is possible to argue that prediction has already served the vital purpose of raising awareness of the kind of problems that BAU might entail. Whether further investment in pursuing ever more precise predictions would now be as incrementally valuable as alternative uses of those resources is an open question.

Sarewitz *et al.* also conclude (2000, 380) that prediction products are least appropriate where there is: a) low or unknown predictive skill; b) little experience in using predictions with the phenomena in question; c) long characteristic time (the period over which predicted events occur); d) at least one alternative to prediction available; and e) high uncertainty in outcomes of alternative decisions.

Geoengineering fails all five of these tests. It fails a) and b) because of the distance of the futures being predicted, the predictive skill of the modellers is unknown, and equally, the experience in using the predictions is very limited. It may well be that the internal consistency of the models is robust but ultimately, this is not a test of the accuracy of their outputs. That can only be tested empirically when the predicted future arrives and since in most cases the predictions are in the nature of 'what if' models, the vast majority of them will never be realised and therefore never tested empirically. Furthermore, the increased precision of the predictions due to the vastly increased computing power deployed in their fabrication, should not be confused with increased accuracy.

The use of prediction to derive geoengineering policy also fails c) because climate change generally and geoengineering in particular have very long characteristic times measured in decades and centuries. Similarly it fails d) because the alternative heuristic approach using systems thinking is available and does not depend on predictions. Finally, it fails e) because the path dependent reflexive nature of interventions at scale in any system that has a material bearing on human (and other biotic) well-being is such that there is inescapable uncertainty about the outcomes of alternative policy decisions, uncertainties that are magnified as the temporal horizon increases providing more opportunity for surprises.

Conclusion

I have argued that prediction is currently at the heart of a predict and control approach to climate policy formulation, but that it has severe limitations as a policymaking tool because it is incapable of coping with the complex dynamics of climate change as demonstrated by reference to Sarewitz *et al.*'s criteria for predictive policymaking. The evidence from the first three papers reviewed in this chapter suggests that using systems thinking is beginning to happen in public policy but as yet is very limited. The five climate change related papers indicate that predict and control remains the overwhelmingly dominant paradigm and although there is evidence in them for systems thinking, it is very slight. The position taken by the recent NAS reports (2015a; 2015b) evidences a positive move towards wider public engagement and experimentation that could be consistent with a systems thinking approach if appropriately executed but does not reframe climate change in terms of complex adaptive systems theory. In the next chapter I will consider how systems thinking and adaptive management might be adopted as a complementary tool by policymakers allowing them to reduce their reliance on prediction and make the paradigm shift away from attempting to solve climate change towards managing it.

Notes

1 Question 332 to the Select Committee on Environment, Transport and Regional Affairs July 1998, accessed 14 March 2013 and available online at www.publications.parliament.uk/pa/cm199899/cmselect/cmenvtra/30/8072209.htm/.

2 I do not wish to convey any sense of criticism of politicians in regard to managing policy changes. This is a complex area and the public and media are probably as much implicated in how such changes are perceived as the politicians themselves.

3 This 'watching brief' status was also confirmed in a personal conversation (19 September 2012) with Jolene Cook at the UK Department of Energy and Climate Change who, at the time, was the civil servant responsible for advising Ministers on geoengineering.
4 The full evidence is available online at www.publications.parliament.uk/pa/cm201012/cmselect/cmenergy/1065/1065vw.pdf (accessed 14 March 2013).
5 Available online and accessed 13 March 2013 at www.publications.parliament.uk/pa/cm201213/cmselect/cmdfence/106/106vw.pdf/.
6 The Oxford Principles demand that geoengineering research be regulated as a public good, with public engagement, transparency of research activities and outcomes, independent assessment, and governance before deployment. Available online at www.geoengineering.ox.ac.uk/oxford-principles/principles/? (accessed 10 December 2012).
7 These activities are undertaken by separate agencies each operating multiple research programmes: National Science Foundation, The National Oceanic and Atmospheric Administration (NOAA), Department of Energy, The National Aeronautics and Space Administration (NASA), Environmental Protection Agency (EPA), Department of Agriculture, Department of Defense and Department of State.
8 The term *global controller* is used with the specific sense given to it by Levin in relation to CAS theory.
9 The common framework comprises eleven criteria – effectiveness, feasibility, scalability, sustainability, environmental risks, costs and affordability, detection and attribution, governance challenges, ethical issues, social acceptability, and uncertainty.
10 The five projects are: a) regional SRM to reverse Arctic warming; b) moderating tropical cyclones with cloud brightening; c) redirecting storms on the US Atlantic coast using cloud brightening; d) compensating for pollution-control induced loss of industrial sulphur aerosols with atmospheric SO_2 injections; and e) slowing ice stream calving using cloud brightening and wave powered vertical mixing.
11 q.v. *Oxford English Dictionary*.
12 More details can be found on their website at http://bipartisanpolicy.org/.
13 *Climate remediation* is the term adopted by the BPC to refer to geoengineering.
14 Private communication with Jolene Cook, Climate and Energy: Science and Analysis (CESA), Department of Energy and Climate Change on 19 September 2012.
15 Data sourced from US Food and Agriculture Organization online at www.fao.org/docrep/w7241e/w7241e05.htm#1.2.1%20photosynthetic%20efficiency (accessed 30 January 2013).
16 As discussed in Chapter 5, Sarewitz *et al.*'s criteria are: a) will predictions take more time to develop than policymakers have to react? b) are accurate predictions necessary or are there viable alternatives? c) can the predictions be understood by non-experts to avoid their misuse or abuse? d) are the criteria for success in scientific prediction similar to those for policymaking to avoid confusion and controversy? e) is there a risk that a focus on prediction will divert resources away from other activities that might better address the issue at hand? (Sarewitz, Pielke, and Byerly 2000, 18–20).

References

Bipartisan Policy Center Task Force on Climate Remediation Research (BPC). 2011. 'Geoengineering: A National Strategic Plan for Research on the Potential Effectiveness, Feasibility, and Consequences of Climate Remediation Technologies'. Bipartisan Policy Center. http://bipartisanpolicy.org/sites/default/files/BPC%20Climate%20Remediation%20Final%20Report.pdf.

Cilliers, Paul. 2002. 'Why We Cannot Know Complex Things Completely'. *Emergence* 4(1/2): 77–84.

Crutzen, Paul J. 2006. 'Albedo Enhancement by Stratospheric Sulfur Injections: A Contribution to Resolve a Policy Dilemma?' *Climatic Change* 77(3–4): 211–220. doi:10.1007/s10584–10006–9101-y.

Gordon, B. 2011. *Engineering the Climate: Research Needs and Strategies for International Cooperation*. DIANE Publishing.

Holling, C. S. 1977. 'The Curious Behavior of Complex Systems: Lessons from Ecology'. In *Futures Research: New Directions*, edited by H. A. Linstone and W. H. C. Simmonds, 114–129. Addison-Wesley Publishing Co. Inc.

House of Commons. 2010. 'The Regulation of Geoengineering'. House of Commons Science and Technology Committee. www.publications.parliament.uk/pa/cm 200910/cmselect/cmsctech/221/221.pdf.

Humphreys, David. 2011. 'Smoke and Mirrors: Some Reflections on the Science and Politics of Geoengineering'. *The Journal of Environment and Development* 20(2): 99.

IPCC. 2012. 'Meeting Report of the Intergovernmental Panel on Climate Change, Expert Meeting on Geoengineering'. IPCC. www.ipcc.ch/publications_and_data/ publications_and_data_supporting_material.shtml#.T9UY98Vr_Uw.

Kuhn, T. S. 1962. *The Structure of Scientific Revolutions* 1. University of Chicago Press.

Lebcir, Mohamed. 2006. 'Health Care Management: The Contribution of Systems Thinking'. University of Hertfordshire Business School Working Papers UHBS 2006–2007. http://uhra.herts.ac.uk/handle/2299/683.

Lempert, R. J., Steven W. Popper, and Steven C. Bankes. 2003. 'Shaping the Next One Hundred Years'. RAND Corporation. www.rand.org/pubs/monograph_rep orts/MR1626.html.

Marshy, Leila, ed. 2010. 'Geopiracy: The Case Against Geoengineering'. ETC Group. www.etcgroup.org/upload/publication/pdf_file/ETC_geopiracy2010_0.pdf.

NAS. 2015a. 'Climate Intervention: Carbon Dioxide Removal and Reliable Sequestration'. National Academy of Sciences. www.nap.edu/catalog/18805/climate-inter vention-carbon-dioxide-removal-and-reliable-sequestration.

NAS. 2015b. 'Climate Intervention: Reflecting Sunlight to Cool Earth'. National Academy of Sciences. www.nap.edu/catalog/18988/climate-intervention-reflecting-sunlight-to-cool-earth.

NERC. 2010. '"Experiment Earth?" Public Have Their Say on Technologies to Reduce Global Warming'. September 9. www.nerc.ac.uk/press/releases/2010/ 35-experiment.asp.

Ricigliano, Robert, and Diana Chigas. 2011. 'Systems Thinking in Conflict Assessment: Concepts and Application'. USAID. http://pdf.usaid.gov/pdf_docs/ PNADY737.pdf.

Sarewitz, Daniel, Roger A. Pielke, and Radford Byerly, eds. 2000. *Prediction: Science, Decision Making and the Future of Nature*. Island Press.

Shepherd, John, Ken Caldeira, P. Cox, J. Haigh, David W. Keith, B. Launder, Georgina Mace, G. MacKerron, J. Pyle, and Steve Rayner. 2009. 'Geoengineering the Climate: Science, Governance and Uncertainty'. The Royal Society.

Stern, N. H. 2007. 'The Economics of Climate Change: The Stern Review'. Cambridge University Press.

Walker, Warren E., Marjolijn Haasnoot, and Jan H. Kwakkel. 2013. 'Adapt or Perish: A Review of Planning Approaches for Adaptation under Deep Uncertainty'. *Sustainability* 5(3): 955–979.

Wigley, T. M. L. 2006. 'A Combined Mitigation/Geoengineering Approach to Climate Stabilization'. *Science* 314(5798): 452–454. doi:10.1126/science.1131728.

7 Reframing geoengineering from solution to contribution

IPCC AR5 demonstrates the growing interest in geoengineering. From cursory or no mentions in the previous four Assessment Reports, AR5 has comprehensive sections on geoengineering in each of the three Working Group reports. This is an unequivocal majority statement by the global climate science community that geoengineering is worthy of consideration. However, while the AR5 Working Group I Summary for Policymakers briefly and dismissively mentions geoengineering, the Summaries for Policymakers of the two working groups concerned with the impacts of and responses to climate change are entirely silent on the subject. This is an equally unequivocal statement that the global policymaking community is not yet ready to embrace geoengineering as a potentially valuable part of the policy mix. This chapter examines the framing of geoengineering, identifying two polarised views, geoengineering as a possible solution to global warming and geoengineering as a complement in a broad-based policy mix. It rejects the 'solution' framing on the grounds of its practical infeasibility and considers from a complex adaptive systems perspective the governance implications of geoengineering as a part of a broad policy mix, opening up a discussion that is continued in the next chapter.

Framing – geoengineering as the solution to our problems

As already noted, framing matters, yet in the geoengineering literature the authors' understanding of how geoengineering fits into the wider climate change discourse is usually not explicit (Bellamy *et al.* 2012). In one of the earliest references to geoengineering in a US government report in 1965, it was presented as the only route to dealing with climate change and the idea of reducing emissions was not even mentioned (Environmental Pollution Panel 1965; Keith 2000). More recently, Crutzen's landmark paper (2006) presents SRM geoengineering as 'the only option' in the face of a failure to abate emissions. Even more recently, Bellamy *et al.*'s survey of the literature shows that geoengineering 'has largely been appraised in contextual isolation, ignoring the wider portfolio of options for tackling climate change [...] and creating an artificial choice between geoengineering proposals'.

Both for those who consider geoengineering to be a means of avoiding aggressive emissions abatement and for those who object to it on ethical grounds, geoengineering is presented as the dominant response to climate change. The American Enterprise Institute, a conservative US think tank, even while still questioning the gravity of climate change, holds that 'if climate change is a significant problem, then SRM could be part of a significant solution' and suggests that the incentives for using SRM appear to be greater than those for emissions abatement (Lane and Bickel 2013).

Most of the early discourse on geoengineering framed it as a potential solution to the climate change problem. From fields of foil a million miles from the Earth or giant orbiting mirrors reflecting away a small proportion of the sun's incoming energy, to vast tracts of new forests and the continuous refilling of evacuated oil and gas wells with sequestered atmospheric CO_2, geoengineering was presented as an alternative to emissions reduction, a means of obviating both the need for extensive and costly adaptation projects and the major lifestyle changes some thought would be the inevitable result of decarbonising the global economy. In this formulation geoengineering would need to be implemented rapidly, globally and at climate-controlling scale. For many it seems that this has been an unwitting framing although others have articulated it as a preferred policy response. ETC (Marshy 2010), a Canadian environmental activist group implacably opposed to geoengineering, refer to the 'Lomborg manoeuver' in reference to Bjørn Lomborg, author of *Cool It: A Skeptical Environmentalist's Guide to Global Warming* (Lomborg 2007) who, they claim, followed other prominent conservatives, such as the American Enterprise Institute, the Cato Institute, and the Heartland Institute, in switching from denying climate change to supporting geoengineering as the dominant response.

This framing continues to be prevalent even amongst those without vested interests or, as in Lomborg's case, those with an unswerving conviction that future generations will be better placed to confront climate change if we invest in economic growth rather than attempt to decarbonise the economy. For example, Lomborg argues that '[r]adically cutting carbon-dioxide emissions will be far more expensive than adapting to higher temperatures' (Lomborg 2001). Hulme (2014) in a book devoted to unmasking SRM as a 'hubristic techno-fix' that is 'undesirable, ungovernable and unreliable' follows others (Fleming 2010; Gardiner 2010; Goodell 2010; Hamilton 2013; Kintisch 2010) in presenting SRM as the dominant response to climate change by which the global surface temperature could be reset, often referred to as a 'thermostat in the sky'. This is the Plan A/Plan B dichotomy that emerged from Crutzen's 2006 paper.

Preston (2011) also draws on Goodell's book *How to Cool the Planet* (2010) in which David Keith is reported as advising caution before 'creating a managed planet'. Preston's argument assumes that geoengineering would deliver a 'managed planet' in the direct sense that humanity could design and intervene in planetary systems in order that they better serve human needs than without such an intervention. Preston's framing of geoengineering as 'a way to modify Earth's parameters so that humans do not need to change' casts it as the dominant response. He does not discuss, other than tangentially, the ethical implications stemming from geoengineering being just part, even possibly a small part, of the policy mix. Nor is it clear that his implicit proposition, that the planet would meekly yield to humanity's design ambitions, is valid. Gunderson and Holling advise caution concerning 'the dangerous myth that the variability of natural systems can be effectively controlled' (Gunderson and Holling 2001, 28). Humanity undoubtedly has the power to intervene at scale but whether it yet has the capacity and wit to avoid creating the climatic equivalent of Frankenstein's Monster, is far from clear (Charlesworth and Okereke 2010).

In the following sections I argue that quite apart from any philosophical or ideological objections to geoengineering as the dominant response to global warming, there is sufficient theoretical understanding of the implications of all forms of geoengineering to be confident that none offers a practical alternative to emissions abatement and that framing them as such creates unfounded and distracting fears of perils from their extensive and long-term deployment that would never arise.

Reliance on SRM

A little reflection on the viability of SRM as the dominant response to climate change and as an alternative to emissions abatement is sufficient to discard it as a credible possibility (Matthews and Caldeira 2007). A policy of continuing unabated carbon emissions whose greenhouse effect was counteracted only by SRM might be extremely effective at constraining GMST within a narrow range not too far from its Holocene values, but would do nothing to prevent considerable climate change. Illustrating the dangers of using global averages as policy targets, modelling studies (Brovkin *et al.* 2009; Irvine, Ridgwell, and Lunt 2010) confirm the intuitively obvious that SRM is not capable of restoring or maintaining the global climate in some desired prior state but would cause radical changes in global weather patterns that would, in turn, cause major and widespread ecosystem impacts, even while stabilising GMST.[1] Moreover, SRM would do nothing to prevent further ocean acidification. How resilient the ocean ecosystem is to further acidification is an open question but there are grave concerns about unabated carbon emissions. IPCC AR5 WGII (Box TS.7) notes that 'mass extinctions in Earth history occurred during much slower rates of change in ocean acidification' and that ocean acidification in high emission scenarios risks '[c]omplex additive, antagonistic, and/or synergistic interactions [...] with disruptive ramifications for ecosystems as well as for important ecosystem goods and services'. Irrespective of the SRM technology used, a predominantly SRM managed climate, far from stabilising the climate, opens a Pandora's box of climate change futures. This climate instability occurs, paradoxically, while GMST is (or might be) stabilised. The impossibility of stabilising the climate by relying solely, or even primarily, on SRM renders ethical challenges to an SRM dominant climate policy largely redundant.

Reliance on CDR

CDR fares little better. In one idealised and extreme modelling study (Cao and Caldeira 2010), it was estimated that in addition to the cessation of all anthropogenic emissions in 2049, almost half the cumulative anthropogenic emissions, amounting to some 1,100PgC by 2049, would also have to be sequestered over the ensuing 80 years in order to stabilise atmospheric concentrations at their pre-industrial level of 278ppm. This study cannot be used to assess the feasibility of relying on CDR to sequester the continuing unabated annual emissions, but it does clearly perform its intended function of illustrating the scale implications of a reliance on CDR and its complex effects on the planetary carbon cycle.

If a CDR dominant policy were contemplated, it is instructive to examine the physical scale of the task. In 2012 CO_2 emissions released some 10PgC into the atmosphere (IPCC 2014: SPM B. 5). To this must be added the non-CO_2 GHGs that are estimated to be equivalent to about 5PgC, yielding an annual total of the order of $15PgC_{eq.}$[2] This compares to current global annual coal production of slightly less than 8Pg.[3] The scale of the task of sequestration is immediately apparent. If CDR geoengineering were to be relied upon to remove all current annual GHG emissions, it would be the equivalent of burying an amount of carbon over twice the mass of globally mined coal for the indefinite future. However, because the carbon density of coal is greater than that of CO_2, the equivalent mass of liquefied CO_2 would be three times that of the corresponding amount of coal and the liquefied CO_2 would occupy over five times its volume.

If CDR were in addition used to reduce current atmospheric GHGs (so-called 'negative emissions') modelling studies (Vichi, Navarra, and Fogli 2013) suggest that as atmospheric CO_2 concentration reduces so CO_2 previously sequestered naturally in the oceans is outgassed, producing an additional burden for CDR. Vichi *et al.* show that the amount of this outgassing is subject to considerable uncertainty but probably reduces the longer the period over which the CDR occurs. An aggressive CDR policy would imply a significant incremental burden from outgassing, whereas if undertaken over several decades, the outgassing might be minimal. To return to the 1990 level of approximately 350ppmv from today's 400ppmv, as proposed by activists such as 350.org, would require the removal of 50ppmv (assuming optimistically that there is no significant ocean outgassing). This represents in excess of 100PgC[4] and if carried out over, say 40 years, the total annual sequestration target, aggregating the current and negative emissions, would start at 18PgC and increase with the growth in annual emissions that, since

2000, has averaged more than 1$PgCyr^{-1}$. If the corresponding amount of CO_2 were liquefied, its volume would be about seventeen times the volume of oil currently extracted globally each year, and by weight, almost ten times the current global annual production of coal. It will be a long time, if ever, before CDR at this scale is feasible as is argued in the next two sections examining the biomass and physical sequestration options.

Biomass sequestration

Since aggregate NPP[5] is of the order of 100$PgCyr^{-1}$ (Field *et al.* 1998), an 18 per cent increase in NPP would be needed to stabilise atmospheric GHGs at their 1990 level. The modest experiments to date in seeding the oceans to increase marine biomass have not suggested that this is likely to be a successful CDR technology, and certainly not at the scale required here. If the burden is to fall entirely on land, because NPP is divided roughly equally between land and oceans (Field *et al.* 1998), an increase of 36 per cent would be required in terrestrial NPP. Moreover, this would need to be done every year into the indefinite future although it could possibly be scaled back slightly once the desired atmospheric concentration had been reached. Given that we have no means of significantly increasing the efficiency of photosynthesis,[6] the only way to achieve this growth is by a combination of increasing the amount of land devoted to biomass, and increasing biomass yields per hectare.

Assuming that we do not want to encroach on the 13 per cent of land already used as cropland, and allowing for the 28 per cent of land already covered by forests, and excluding the 28 per cent that is not cultivatable (deserts, snow and glaciers, bare soil, built land (cities, quarries, etc.), and inland water bodies), the incremental NPP must be created in the remaining 31 per cent currently comprising grassland, shrub and herbaceous land and sparsely vegetated land, that in most cases will be of relatively low fertility or accessibility.[7] If we suppose that half of this is cultivatable, the 18 per cent increase in NPP must be produced on just 15 per cent of the planet's land surface, or less than 5 per cent of its total surface. Even if this were possible, it would require massive technological interventions. Interventions at this scale are not yet available and even if they were, would almost certainly have side effects elsewhere in the complex web of relationships that sustain humanity and its fellow biota that would need to be carefully researched before embarking on such a grand reshaping of the planet's surface.

The demands of scale to make biomass CDR capable of sequestering a major part of the 18$PgCyr^{-1}$ make it infeasible as an alternative to emissions abatement. Moreover, even it were feasible, because of the low permanence of most forms of biomass sequestration,[8] a biomass sequestration dominant climate policy would bequeath an ever-increasing challenge to future generations. This would entirely confound the demands of intergenerational justice that motivate our concern about climate change.

Physical sequestration

The feasibility of physical sequestration of multiple petagrams of CO_2 suffers from the same resource limitations highlighted by Fox in relation to building emissions abatement and adaptation infrastructure, and other constraints discussed in Chapter 2. 18PgC in the form of liquefied CO_2 would represent 17 times the current annual volume of oil production. For illustrative purpose, I assume the sequestration is done by DACS. Given that atmospheric CO_2 is a trace gas at ~400ppmv, and the 'capture fraction', that proportion of the CO_2 in a given volume of air that is extracted during one pass across the DAC contactors, is about 50 per cent (Holmes and Keith 2012), the amount of air that would need to be processed to extract 18PgC would be about 4 per cent of the Earth's entire atmosphere every year. This ignores any offsetting emissions, upstream in building the DACS infrastructure and in the capture and compression process, and downstream in sequestering the liquefied CO_2 and keeping it sequestered indefinitely. Based on Holmes and Keith's estimates that one DAC unit could produce 76ktCyr^{-1}, sequestering 18PgCyr^{-1} implies almost 200,000 such units at a capital cost of $160bn. Keith does not consider this to be feasible.[9]

In each of these calculations I have assumed that geoengineering is the dominant response to global warming, that emissions abatement would be a low priority, that adaptation would be entirely reactive, and that the task is to remove the entire excess CO_2 from current and historical emissions. The conclusion is that this is unrealistic. However, even if several of these technologies were used, each to provide, say 1PgCyr^{-1} of sequestration, and only half the emissions were sequestered, the aggregate resource demands for CDR would still entail an extraordinary diversion of materials, manpower and finance from other activities more focussed on providing goods than removing bads. They would create new industries at global scale that would be ecologically intrusive at a time when we are trying to reduce our planetary footprint, and almost certainly be the source of new risks in a series of significant turns of the Beckian reflexive modernisation screw.

In short, critiques of geoengineering that proceed from it being the principal or dominant response should be discarded as they rest on the fundamentally flawed initial premise that some combination of SRM and CDR geoengineering technologies could feasibly offer a practical alternative to emissions abatement and adaptation as a long-term effective response to climate change. However, other framings are available.

Framing – geoengineering as an emergency response

Partly stemming from Crutzen's proposition that geoengineering might be needed because of the ineffectiveness of emissions reduction policies, geoengineering is often presented as an emergency option (Bellamy *et al.* 2012). This framing sees geoengineering as analogous to a lifeboat to be deployed in a crisis. There are several problems with this framing. First, if it is not to be deployed until it is used in earnest, by definition it cannot be properly tested prior to full deployment. Without prior testing at scale, it could not be known whether geoengineering in any of its many guises could be an effective emergency response. How quickly could the geoengineering be deployed? What would be the side effects of a rapid emergency scale geoengineering intervention in the global climate? Would there be sufficient time for the geoengineering to have the necessary climatic effects before the crisis became a catastrophe? Readiness time, deployment time, impact time all become critical risk factors and in the absence of any empirical testing they would be subject to great uncertainty. Heroic and untried interventions in complex systems are likely to have significant unintended consequences. Many of these are likely to be highly undesirable because major interventions generally cause major perturbations and these test, perhaps to destruction, the resilience of the systems so perturbed.

The lifeboat analogy has additional limitations. First, whereas a lifeboat is a boat and there is a great deal of human experience with boats that gives high confidence about how a lifeboat will operate in an emergency situation, geoengineering is both novel and *sui generis*, and there is no accumulated empirical knowledge about how it might work or what its systemic side effects might be. Furthermore, a lifeboat is a local response and the failure of any particular lifeboat would not have systemic repercussions. This would not be so for a global scale emergency deployment of geoengineering.

Second, the definition of a crisis or emergency sufficiently grave to trigger a geoengineering emergency response is likely to be highly contentious. Hulme (2014) discusses this question at some length concluding that '[c]alling down climate emergencies to justify radical techno-fixes [...] may be an attractive political strategy for some, but carries considerable risks'. Third, as already noted, geoengineering conceived only as an emergency response must perforce be untested prior to deployment. There is only one climate system and no amount of modelling can establish without unquantifiable uncertainty what the medium to long-term consequences, both direct and indirect, of a large scale geoengineering deployment would be. This can only be done empirically but as Hulme observes, there is no miniature Earth on which to experiment. For these reasons, deploying geoengineering as an emergency response brings to mind the instructions on domestic fireworks – light the blue touch paper and stand back – but at global scale. It is hard to conceive of a more high risk and foolhardy strategy than framing geoengineering only as an emergency response to avert a supposed imminent climate catastrophe.

Framing – geoengineering as an adjunct

If geoengineering is not to be the dominant response to climate change, nor an emergency response, does it have a role in a mix of policies? The Plan A/ Plan B dichotomy – emissions abatement is not working, we have to rely on geoengineering – has been rejected. For example, Keith and MacMartin (2015) argue, specifically in relation to SRM, that its risks and benefits depend upon assumptions about the scale and nature of its implementation and they propose an implementation scenario that is far from geoengineering being the dominant policy tool. MacCracken (2006; 2008; 2009a; 2009b; MacCracken in IPCC 2012, 55) has consistently argued that geoengineering should be framed as an adjunct rather than an alternative to emissions reduction and adaptation.

> Basically, the question that needs to be considered is not whether climate engineering on its own would be beneficial or detrimental for the environment and society, but whether society and the environment would be better off working through the consequences of eventually controlling greenhouse gas emissions with or without the partial counterbalancing of at least some of the consequences using climate engineering technologies.
>
> (MacCracken in IPCC 2012, 55)

Others, notably Wigley (2006), have framed geoengineering as a means of 'buying time' for emissions reductions and adaptation to take effect. In this formulation, geoengineering is started early and only at sufficient scale to slow the onset of climate change while concerted efforts are made to decarbonise the global economy. Fox (2009) provided an additional justification for this framing by arguing that there were insufficient human and material resources to build the emissions reduction and adaptation infrastructure necessary for a timely and effective response to climate change and therefore geoengineering was needed to bridge the gap. He refers to this as the MAG approach – combining the initial letters of mitigation, adaptation and geoengineering. In the 'buying time' framing, geoengineering would gradually be reduced to zero as the other responses began to take effect, also a feature of Keith and MacMartin's scenario.

The 2015 NAS reports on 'climate intervention' are explicit that emissions abatement must be the core response to climate change but that it may also be necessary to include some forms of climate intervention in the policy portfolio (NAS 2015a; NAS 2015b), because, as they summarise:

> With whatever portfolio of technologies the transition is achieved, eliminating the carbon dioxide emissions from the global energy and transportation systems will pose an enormous technical, economic, and social challenge that will likely take decades of concerted effort to achieve.
>
> (NAS 2015b)

This framing is consistent with a systems thinking approach because it abandons reductionist notions of solving the climate problem in favour of seeking ways of incrementally improving the situation. These improvements can come from a wide range of interventions some of which are intended simply to stop making matters worse, while others might be more focussed on making things better; geoengineering can contribute to both. What specific geoengineering technologies, at what specific scale and timing would be most suitable can only be determined heuristically. There is such a wide range of possibilities that not all can be researched. However, a systemic approach does not require comprehensive examination of every possibility, rather it demands only that sufficient are explored to allow a continual flow of incremental advances that over time accumulate to achieve the desired outcome of averting dangerous climate change. It follows that research policies that tend to increase the range of potential geoengineering contributions to the policy mix will be preferable to those that tend to limit it. This imperative has direct implications for the governance of geoengineering research.

The complexity approach to governance

The core principles of complex adaptive systems are that changes in them are driven by non-teleological adaptations that are autonomously selected by virtue of their empirically proven superiority in enhancing the system's survival and flourishing in the reality of its changing environment. This is achieved by balancing the often competing demands of resilience and growth by promoting diversity both in the elements in the system and their interactions, that in turn increases the potential for the emergence of novel adaptations from which those that best serve the system's changing needs can be selected and further enhanced. Inimical to the flourishing of an adaptive system is the system-wide imposition by a global controller of its chosen innovations. Not only are these likely to serve the interests of the global controller more than the well-being of the system as a whole, but they are also likely to entail greater risk to the integrity of the system because the consequences of any catastrophic failure are considerably more likely to be systemic than local.

For geoengineering research, the demands of diversity and the suppression of the emergence of a global controller suggest a deliberate policy not to create a formal top-down global governance regime that decides who can do what, how and when, but rather to encourage one that is bottom-up and therefore likely to be more culturally and intellectually diverse. If this diversity is to enjoy the greatest opportunity for serendipitous discovery it must conform to the third of the Oxford Principles for geoengineering research (Rayner *et al.* 2013), namely that '[t]here should be complete disclosure of research plans and open publication of results'. This transparency is essential if the process of selection and enhancement at the root of complex adaptivity is to function effectively (Jasanoff 2010). Onora O'Neill, in a public lecture about trust and geoengineering,[10] cautions that transparency may not be enough, it is also necessary to communicate proactively. Simply placing information in the public domain may, she suggested, sometimes be as much about rendering it impotent as it is about promoting active debate.

There remains the question of risk management. For risk to be controlled requires the scale of innovation to be limited but not to such an extent that experimental results are lost in the noise of natural variability. Complex systems control the potential for catastrophic innovation by limiting to the local, the multiplicity and diversity of the interactions that produce emergent properties. This has the significant effect of reducing the likelihood that the consequences of any failure would be systemic; it may be catastrophic locally but not systemically, and complex systems routinely absorb local catastrophes.

Ubiquitous digital communications have fundamentally altered the spatial dynamics of localness in complex human systems. This applies equally to the geoengineering community. In a globalised world, with all geoengineering actors connected by virtually instantaneous communications, they constitute a globalised 'local' community. Consequently, it is entirely plausible that this community, if left to its own devices, might succumb to groupthink allowing it to act, albeit unwittingly, as a global controller, thereby suppressing diversity of innovation and constraining the emergence of more apt enhancements.

Despite the vibrant debate within the academy on the merits and demerits of geoengineering, the selection by the elite of the participants in key parts of this process already illustrates this suppression of diversity. For example, the recent NAS report on geoengineering was prepared by a panel exclusively comprising natural scientists (NAS 2015a). Admittedly they included James Fleming and Ray Pierrehumbert, the former an atmospheric scientist turned historian and the latter a professor of geophysical science, both known for their sceptical views on geoengineering, but there were no social scientists able to challenge assertions such as CDR not introducing novel global risks, a view based on a narrow climatic assessment of these technologies that ignores their potential environmental and social impacts when undertaken at climate-changing scale. There were no ethicists addressing the moral issues raised by geoengineering. There were no members of civil society, the media or the policymaking community to bring their perspectives to the NAS's deliberations. The more the geoengineering community treats geoengineering as primarily a natural science issue, the more it isolates itself from wider public engagement, and the greater the danger that its policy influence could become dominated by an ever smaller number of voices from within the scientific elite. Indeed, the NAS reports acknowledge this in recommendations that geoengineering governance should reflect the views of 'a broad set of stakeholders'. However, it is not sufficient for the academy to publish its reports and look passively for the engagement of the wider community, the demands of diversity require, much as Onora O'Neill suggests, that they actively engage that wider community in their deliberations.

The governance structure that emerges from these considerations must both encourage innovation by being open and diverse, and control risk by being local and limiting scale. Academic researchers have a long and successful record of controlling experimental risks. They have a structure of peer review to ensure the scientific integrity of research proposals, and in addition have processes of both financial and ethical review that engage those from different disciplines whose role it is to ensure the acceptability of a project's wider risks, whether to the research institution itself or beyond to the research subjects and the public. These processes do not provide certainty against experiments causing unintended and material damage to innocent victims, but this so rarely happens that these processes are demonstrably successful at controlling these risks. Moreover, even when they do arise they are local rather than systemic in their impact. There is no instance of which I am aware, of a formal academic research project ever having caused a pan-global catastrophe. There is no reason to assume that, uniquely, geoengineering research would provoke such an outcome provided it followed long-established and effective research protocols.

No geoengineering technology relies on physics or chemistry that is novel; the risks with all forms of geoengineering derive entirely from their proposed scale and are not limited to their climatic impacts. Indeed, most forms of CDR would entail considerable social and ecological risks if undertaken at climate changing scale even though their direct climatic impacts are thought to be relatively benign. These risks are rarely acknowledged. For example, the NAS report (2015a) describes CDR as 'generally of lower risk and of almost certain benefit' and constrained only by 'cost and lack of technical maturity'. But there will be a threshold below which all geoengineering technologies will be harmless. In climatic terms, either they will have no climate effect or it will be small enough to be rapidly buffered by the climate system much as a wave removes footprints from a sandy beach.[11] Similarly, beneath that threshold, their social and ecological impacts will be insignificant.

The benign image of waves washing away footprints is not matched by the horror of a crevasse opening up underfoot as one steps lightly across a polar landscape. The challenge for geoengineering governance is to allow the technologies to be researched, and if and only if then deemed appropriate, developed and deployed, without stepping at any stage on a metaphorical crevasse. If the boundary between research activities that are benign and those that entail unacceptable risks can be managed, geoengineering may have a role to play in a mixed portfolio of responses to climate change. We have established that this will not be a dominant role, but its significance in that mix can only be determined heuristically.

One way to reduce the risk of a catastrophe is to prohibit action, or so severely constrain it as to render it inconsequential. This option exists for geoengineering and has already been recommended by the UNEP Convention on Biodiversity that in 2010 proposed a moratorium on all forms of geoengineering that may affect biodiversity.[12] However, a moratorium is a blunt instrument that may, for a short period, be valuable as a restraint in certain circumstances where for example a new governance regime is on the threshold of being implemented, or certain imminent research results are required to inform further research. But moratoria risk stifling potentially valuable creativity if deployed inappropriately.

Geoengineering research is where aviation was a century ago. The governance challenge is how to develop policies that embrace the notion of safe-fail, that allow whatever benefits may be achievable from geoengineering to emerge in such a way that any failures along the way can be accommodated. Many people have died from aviation accidents due to technical and operational failures or weather related crashes, those test flying prototypes, innocent passengers and those on the ground merely unfortunate enough to have been in the wrong place at the wrong time. As I edit this chapter, the news is unfolding of the tragedy of the Germanwings aircraft deliberately crashed by its co-pilot into the French Alps, turning his suicide into the murder of the other 149 people on board. Nevertheless, flying remains one of the safest forms of travel and a central component of the globalised world in which we live. Despite many accidents being catastrophic for those involved, those catastrophes were local. At a global level, aviation is regarded as an extraordinary success story. Systemic flourishing and resilience has been achieved, at least in part, at the price of local catastrophe. This is not to suggest any degree of complacency about risk but rather to highlight five key risk aspects of innovation. First, without some risk there can be no innovation. Second, what constitutes an acceptable risk is normative and relative. Third, while every hazard may be avoidable, the occurrence of some hazards is certain – improbable events are just that, improbable, not impossible. Fourth, the damages suffered when a risk crystallises cannot always be reversed or compensated for, but this is not a sufficient reason not to incur the risk if it is done in a measured and responsible way. Last, not all risks can be foreseen, the future is pregnant with surprise.

Systems designed to fail safely are much less likely to cause systemic catastrophes than those designed not to fail. Safe-fail in a designed system requires structural constraints to ensure that failures are locally contained, even though they may be locally catastrophic. In designing experiments this implies two safety mechanisms; first the scale of any single failure must be small enough that it does not directly threaten the survival of the system. Second, the equivalent of firebreaks are needed to ensure that a modest local failure does not become a systemic failure by being allowed to cascade across the system. For geoengineering, the first is achievable by limiting the scale of any research project so that it does not encroach too far into territory that remains empirically uncharted. The second is more challenging and probably infeasible. The complexity of the ecosphere is such that it is inconceivable that all the potential domino effects that might arise from a local failure of any human activity can be predetermined, or, even if they were, that structures could be put in place to prevent them. To avoid compromising systemic resilience, reliance must be placed on limiting the scale of any project so that the then boundaries of knowledge are pushed only slightly. But pushed they must be, for if we operate only within the boundaries of existing knowledge we learn nothing new.

The challenge of controlling systemic risk through limiting the scale of experiments does require some global cooperation although it does not necessarily point to the need for a formalised top-down global governance regime. Parson and Keith (2013) have proposed a scale limit as one element of a package to 'end the deadlock on geoengineering research'. The problem with defining the limit as an absolute number, in their case limiting the radiative forcing from any experiment to less than $10^{-2}Wm^{-2}$, is that the perceived risk threshold changes reflexively with our incremental knowledge. How then is this global limit to be reset as circumstances allow, if not by a globally agreed governance regime? I contend that the limit itself can be an emergent property of the diverse and complex system that is the geoengineering community. Importantly, in this construction, this community is no longer restricted to the academic elite at the leading edge of geoengineering research but also includes policymakers, the media, civil society and even the general public, to the extent that they wish to engage with the process. Modern forms of communication allow virtual communities to form with relative ease and largely unconstrained by geographic dispersal. A discussion in such a forum could quickly establish broad parameters for overarching limits for geoengineering research projects. I refer to this forum as the Geoengineering Governance Network (GGN) and it is the subject of the next chapter.

Governance as an obstacle to progress

Chapter 4 of the NAS report on SRM[13] (NAS 2015b) provides a comprehensive review of the literature on geoengineering governance. Drawing on this literature, in their own recommendations they highlight the need for transparency, the engagement of a broad set of stakeholders extending beyond the academy; the need for the governance to respond reflexively to future increases in scope and scale of SRM research, and they also add that 'the goal of governance should be to maximize the benefits of research while minimizing the risks'. They do not take it upon themselves to make specific proposals for governance structures that might deliver these and their other governance objectives.

This framing is consistent with the principles of CAS theory – multiple actors interacting independently promoting the more advantageous emergent innovations in preference to the less favourable. However, the atavistic ties to reductionist thinking remain evident. They assert that:

> It is important to give careful thought to the mechanisms for governing research on albedo modification, since they may later form part of the basis for a mechanism for governing sanctioned or unsanctioned deployment should a choice ever be made to proceed to that stage.

They then use this assertion as a justification for linking the governance of research to the governance of deployment and pose three question they consider will be important in reaching any international agreement about SRM:

1 How is it decided when the benefit to albedo modification will outweigh the harm? What metric should be used?
2 What obligation do the acting parties have to compensate others for damages, anticipated or otherwise, caused by albedo modification? Who decides causality and how is it determined?
3 Who decides what is benefit versus harm, and on what time and space scales are such determinations made?

Two issues need examining. First, is it appropriate to link research and deployment governance in this way, and second, if it is, what is the significance of applying these three questions to the governance of research. Before examining these two issues, it is important to note that the NAS questions, bringing together the nature of and the responsibility for the negative consequences of albedo modification, are expressed in terms that could equally apply to almost any public policy intervention. To illustrate this, I invite you to reread the three questions substituting for 'albedo modification', in turn, 'nuclear power generation', 'press freedom', 'gun control', 'continued GHG emissions' and 'continued economic growth'. These are just five examples selected from a virtually endless list of contested public policy issues. Whereas the first three are policies largely within the domain of sovereign states and the last two are more international in character, the questions formed by their substitution are all equally intractable. As a general rule, policymakers do not insist upon such questions having definitive answers before proposing and then implementing their chosen policy interventions. Indeed, the art of policymaking in socially progressive and responsible societies resides precisely in policymakers exercising judgement and decisiveness even in the absence of broadly accepted answers to unanswerable questions such as these.

For the linking of research and deployment governance to be helpful requires that the criteria that apply in both cases be broadly similar. Yet the argument from complexity would suggest otherwise. The central question is the relationship between scale and risk. Self-evidently, larger scale experiments entail greater risk than smaller ones where they are both based on the same corpus of knowledge. However, where outcomes from small scale experiments inform the design of later larger scale ones, it may be assumed that unless the experimenters are either ignorant or reckless, the degree of risk will be no greater, and will probably be less despite the increase in scale. The example of aviation illustrates that one of the principal purposes of experimentation with new technologies is that risk should not increase with scale.

The modern world is replete with technological innovations that in earlier incarnations would have been considered too risky to contemplate. Flying

400 people for more than 12 hours, half way around the planet, several miles above the Earth's surface in a winged metal tube powered by expelling super-hot gases is about as far removed from the Wright brothers' first flights as it is possible to be and yet be able to describe both as powered flight. Nevertheless, modern civil and military aviation have only become possible because of countless innovations that have systematically reduced the risk of flying. One might ask whether aviation would have got off the ground had the pioneers been obliged to wait until a global system of regulation had been implemented that catered for all the conceivable aviation developments several decades into the future. This is particularly so if one retrospectively applies today's heightened sensitivity to risk to their state of knowledge a century ago, and thereby justifies imposing restrictions that would perhaps have prohibited flights lasting more than five minutes, or flying higher than two metres, or that had the remotest possibility of crashing onto populated areas, or crossing national boundaries. And then, with each innovation, would a global governance entity have revised those limits in a timely fashion to reflect the reduction in risk from the acquisition of new knowledge only made possible by those experiments?

Issues of scale in terms of geographic extent, duration and intensity will be very different in the early stages of research from those applicable to deployment. Indeed, the longer the research and development period, the greater the number of incremental steps and the lower will be the correlation between early risk and later scale. Because each research project informs later ones, the research activity becomes part of a larger complex adaptive system in which predicting the trajectory of yet-to-emerge innovations can be little more than a matter of conjecture. On practical grounds it would seem most inappropriate to assume that the key governance questions applying to early stage research bear much resemblance to those that might be necessary in relation to full scale deployment in the distant future. This is so even though there may be a time when the line dividing research and deployment becomes fuzzy; this point may never arrive and even if it does, it is still a long way off and we may assume that those in the future will be more than capable of recognising it and dealing with it in the most appropriate manner. I will return to our relations with future generations in the final chapter.

The second question concerns the significance of applying the NAS governance questions to geoengineering research, even assuming that there is some correlation between scale and risk. The difficulty arises because the questions are unanswerable. They provoke needed reflection on our relationship with the planet and each other, but they are unanswerable in anything other than the most general terms, and those answers will be culturally specific and contested. These are questions for philosophers whose role is to explore not to act, rather than for policymakers whose task is to exercise judgement in the face of incomplete knowledge in order to deliver practical policies that on balance are more rather than less likely to benefit those to whom they acknowledge some moral and political obligation.

Making small scale empirical geoengineering research contingent upon a prior international agreement on questions such as these creates an impenetrable barrier to progress greatly increasing the costs of inaction, whether financial or human. In complex adaptive systems terms, the assumption that there are settled answers to these questions implies the existence of a global controller and demonstrates its destabilising effect. By imposing its writ across the entire global geoengineering research effort, an institution that purports to speak with one voice in response to these questions increases the risk of catastrophic failure by limiting the possibilities for the emergence of novel processes that would increase systemic resilience and thereby reduce overall risk.

An analogy could be drawn with a couple basing a decision to have a child on their assessment of whether, over its entire life, the child would do more good than harm, the extent to which they would be liable for any harm it might cause and who would be liable after their death, and what constitutes harm in a situation where, for example, the child harms someone in order to benefit another, much as politicians do when taking their countries to war or they tax the rich to provide welfare for the poor. These are important ethical questions precisely because they do not have neat timeless answers. Their purpose is to provoke not to resolve. The governance of geoengineering certainly needs to reflect on these questions and allow those reflections to inform its decisions, but demanding answers to unanswerable questions as a prior condition to conducting research that could be undertaken, if not without risk, then at least with risks that are orders of magnitude less than the risks from not undertaking the research, would be tantamount to institutionalised inaction and seem to come close to the folly feared by Gray and Tuchman.[14]

Conclusion

Geoengineering can only be deployed as a credible response to climate change as part of an integrated policy mix in which it does not play a dominant role. Whether geoengineering can in fact enhance humanity's response to climate change by being part of the policy mix will not emerge until empirical research is undertaken, and if it can, how it can will be subject to continual enhancement with the benefit of experience. Today's governance of geoengineering research should not attempt to integrate highly unpredictable distant future concerns about the control of full scale deployment, for fear of undermining the urgency and diversity essential to the research effort. Research governance needs to be adaptive and must allow global agreement to emerge at the appropriate time rather than forcing it prematurely into existence. In the next chapter I propose a route whereby this might be achieved.

Notes

1 Planetary scale climate interventions would change the spatial distribution of the fundamental physical factors that create the weather (temperature, pressure,

precipitation and so on). These changes would create novel patterns of weather even while GMST was maintained at Holocene levels.

2 Data from the European Commission Joint Research Centre, available online at http://edgar.jrc.ec.europa.eu/ (accessed 3 February 2014).

3 2012 data from the World Coal Association, available online at www.worldcoal.org/resources/coal-statistics/ (accessed 23 September 2015).

4 Each 1ppmv of atmospheric CO_2 amounts to approximately 2PgC or almost 8PgCO_2. Because of the two atoms of oxygen, sequestering carbon in the form of CO_2 implies 3.67 times the mass of material required to sequester it in the form of pure carbon.

5 Primary production is the synthesis of organic carbon compounds from atmospheric or aqueous carbon dioxide. Net primary production (NPP) is the rate at which all the plants in an ecosystem produce net useful chemical energy; it is equal to the difference between the rate at which they produce useful chemical energy and the rate at which they use some of that energy during respiration. It is measured in terms of the mass of carbon sequestered per annum.

6 Certain types of SRM, such as stratospheric aerosol injection (SAI), would have a side effect of increasing the amount of diffused light reaching the Earth's surface and this would promote photosynthesis. However, in the absence of any research that shows that the overall risk/reward profile is improved by deploying SAI in order to improve CDR efficacy, it would be prudent to regard this option with some circumspection.

7 Data from the Food and Agriculture Organization of the United Nations Statistics Division, available online at www.fao.org/news/story/en/item/216144/icode/ (accessed 26 March 2015).

8 See Chapter 2.

9 Personal correspondence.

10 A lecture entitled *Climate engineering: who can we trust?* held on 13 March 2015 as part of the University of Cambridge Science Festival.

11 Even the International Space Station and the countless satellites now orbiting the Earth reflect away some solar radiation but this is insufficient to raise concerns about their climatic effect.

12 The text of the moratorium allows 'small scale research studies' but subject to certain criteria that would be difficult to comply with in the case of geoengineering, not the least of which is that the research has no negative transboundary effects. Researchers in the live environment cannot know *a priori* that their experiments will have no negative transboundary effects. They can limit them by the scale and duration of the experiment but if the purpose of research is to push the boundaries of knowledge, they cannot entirely avoid such risks. This then becomes a question of materiality and that is a normative matter that could be accommodated by negotiation.

13 The NAS reports adopt the term 'albedo modification' that they consider better describes the range of technologies previously categorised as SRM.

14 See references in Chapter 1.

References

Bellamy, Rob, Jason Chilvers, Naomi E. Vaughan, and Timothy M. Lenton. 2012. 'A Review of Climate Geoengineering Appraisals'. *Wiley Interdisciplinary Reviews: Climate Change* 3(6): 597–615.

Brovkin, Victor, Vladimir Petoukhov, Martin Claussen, Eva Bauer, David Archer, and Carlo Jaeger. 2009. 'Geoengineering Climate by Stratospheric Sulfur

Injections: Earth System Vulnerability to Technological Failure'. *Climatic Change* 92(3): 243–259. doi:10.1007/s10584–10008–9490–9491.

Cao, Long, and Ken Caldeira. 2010. 'Atmospheric Carbon Dioxide Removal: Long-Term Consequences and Commitment'. *Environmental Research Letters* 5(2): 024011.

Charlesworth, Mark, and Chukwumerije Okereke. 2010. 'Policy Responses to Rapid Climate Change: An Epistemological Critique of Dominant Approaches'. *Global Environmental Change* 20(1): 121–129. doi:10.1016/j.gloenvcha.2009.09.001.

Crutzen, Paul J. 2006. 'Albedo Enhancement by Stratospheric Sulfur Injections: A Contribution to Resolve a Policy Dilemma?'. *Climatic Change* 77(3–4): 211–220. doi:10.1007/s10584–10006–9101-y.

Environ PollutPanel. 1965. 'Restoring the Quality of Our Environment'. President's Science Advisory Committee. http://dge.stanford.edu/labs/caldeiralab/Caldeira%20downloads/PSAC,%201965,%20Restoring%20the%20Quality%20of%20Our%20Environment.pdf.

Field, Christopher B., Michael J. Behrenfeld, James T. Randerson, and Paul Falkowski. 1998. 'Primary Production of the Biosphere: Integrating Terrestrial and Oceanic Components'. *Science* 281(5374): 237–240. doi:10.1126/science.281.5374.237.

Fleming, James Rodger. 2010. *Fixing the Sky: The Checkered History of Weather and Climate Control*. Columbia University Press.

Fox, Tim. 2009. 'Climate Change: Have We Lost the Battle?'. Institution of Mechanical Engineers. www.imeche.org/NR/rdonlyres/77CDE5E4-CE41-4F2C-A706-A630569EE486/0/IMechE_MAG_Report.PDF.

Gardiner, Stephen. 2010. 'Is "Arming the Future" with Geoengineering Really the Lesser Evil?'. In *Climate Ethics: Essential Readings*, edited by Stephen Gardiner, Simon Caney, Dale Jamieson, and Henry Shue, 285–312. OUP USA.

Goodell, Jeff. 2010. *How to Cool the Planet*. Houghton Mifflin.

Gunderson, L. H., and C. S. Holling, eds. 2001. *Panarchy: Understanding Transformations in Human and Natural Systems*. Island Press.

Hamilton, Clive. 2013. *Earthmasters: The Dawn of the Age of Climate Engineering*. Yale University Press.

Holmes, Geoffrey, and David W. Keith. 2012. 'An Air–liquid Contactor for Large-Scale Capture of CO2 from Air'. *Philosophical Transactions of the Royal Society A: Mathematical, Physical and Engineering Sciences* 370(1974): 4380–4403. doi:10.1098/rsta.2012.0137.

Hulme, Mike. 2014. *Can Science Fix Climate Change?: A Case Against Climate Engineering*. First edition. Polity Press.

IPCC. 2012. 'Meeting Report of the Intergovernmental Panel on Climate Change Expert Meeting on Geoengineering'. IPCC. www.ipcc.ch/publications_and_data/publications_and_data_supporting_material.shtml#.T9UY98Vr_Uw.

IPCC. 2014. *Climate Change 2013 – The Physical Science Basis: Working Group I Contribution to the Fifth Assessment Report of the Intergovernmental Panel on Climate Change*. Cambridge University Press.

Irvine, Peter J., Andy Ridgwell, and Daniel J. Lunt. 2010. 'Assessing the Regional Disparities in Geoengineering Impacts'. *Geophysical Research Letters* 37 (September): 6 pp. doi:201010.1029/2010GL044447.

Jasanoff, Sheila. 2010. 'A New Climate for Society'. *Theory, Culture & Society* 27(2–3): 233–253.

Keith, David W. 2000. 'Geoengineering the Climate: History and Prospect'. *Annual Review of Energy & the Environment* 25(1): 245.

Keith, David W., and Douglas G. MacMartin. 2015. 'A Temporary, Moderate, and Responsive Scenario for Solar Geoengineering'. *Nature Climate Change*, March. doi:10.1038/nclimate2493.

Kintisch, Eli. 2010. *Hack the Planet: Science's Best Hope – or Worst Nightmare – for Averting Climate Catastrophe*. John Wiley & Sons.

Lane, Lee, and J. Eric Bickel. 2013. 'Solar Radiation Management: An Evolving Climate Policy Option'. American Enterprise Institute.

Lomborg, Bjørn. 2001. 'The Truth about the Environment'. *The Economist* 4: 63–65.

Lomborg, Bjørn. 2007. *Cool It: The Skeptical Environmentalist's Guide to Global Warming*. Cyan and Marshall Cavendish.

MacCracken, M. C. 2006. 'Geoengineering: Worthy of Cautious Evaluation?' *Climatic Change* 77(3): 235–243.

MacCracken, M. 2008. 'Geoengineering: Getting a Start on a Possible Insurance Policy'. In *International Seminar on Nuclear War and Planetary Emergencies – 40th Session*, 747–781. www.worldscientific.com/doi/abs/10.1142/9789814289139_0072.

MacCracken, M. 2009a. 'Beyond Mitigation: Potential Options for Counter-Balancing the Climatic and Environmental Consequences of the Rising Concentrations of Greenhouse Gases'. Policy Research Working Paper Series. World Bank. http://papers.ssrn.com/sol3/papers.cfm?abstract_id=1407956.

MacCracken, M. 2009b. 'On the Possible Use of Geoengineering to Moderate Specific Climate Change Impacts'. *Environmental Research Letters* 4(4): 045107. doi:10.1088/1748–9326/4/4/045107.

Marshy, Leila, ed. 2010. 'Geopiracy: The Case Against Geoengineering'. ETC Group. www.etcgroup.org/upload/publication/pdf_file/ETC_geopiracy2010_0.pdf.

Matthews, H. D., and Ken Caldeira. 2007. 'Transient Climate–carbon Simulations of Planetary Geoengineering'. *Proceedings of the National Academy of Sciences* 104(24): 9949–9954. doi:10.1073/pnas.0700419104.

NAS. 2015a. 'Climate Intervention: Carbon Dioxide Removal and Reliable Sequestration'. National Academy of Sciences. www.nap.edu/catalog/18805/climate-intervention-carbon-dioxide-removal-and-reliable-sequestration.

NAS. 2015b. 'Climate Intervention: Reflecting Sunlight to Cool Earth'. National Academy of Sciences. www.nap.edu/catalog/18988/climate-intervention-reflecting-sunlight-to-cool-earth.

Parson, Edward A., and David W. Keith. 2013. 'End the Deadlock on Governance of Geoengineering Research'. *Science* 339(6125): 1278–1279. doi:10.1126/science.1232527.

Preston, Christopher J. 2011. 'Re-Thinking the Unthinkable: Environmental Ethics and the Presumptive Argument Against Geoengineering'. *Environmental Values* 20 (November): 457–479. doi:10.3197/096327111X13150367351212.

Rayner, Steve, Clare Heyward, Tim Kruger, Nick Pidgeon, Catherine Redgwell, and Julian Savulescu. 2013. 'The Oxford Principles'. *Climatic Change* 121(3): 499–512. doi:10.1007/s10584–10012–0675–0672.

Vichi, M., A. Navarra, and P. G. Fogli. 2013. 'Adjustment of the Natural Ocean Carbon Cycle to Negative Emission Rates'. *Climatic Change* 118(1): 105–118. doi:10.1007/s10584–10012–0677–0.

Wigley, T. M. L. 2006. 'A Combined Mitigation/Geoengineering Approach to Climate Stabilization'. *Science* 314(5798): 452–454. doi:10.1126/science.1131728.

8 Geoengineering Governance Network (GGN)

This chapter provides some initial thoughts on the GGN. It is not my intention to present a fully fledged governance regime. If, as I argue, the governance regime is itself a complex adaptive system it must emerge through multiple interactions from which the more apt innovations are selected and promoted for further enhancement. My contribution here can at best be just one pebble in that pond. There are therefore likely to be many objections of both practice and principle that need to be articulated and negotiated. Nevertheless, if the GGN concept is to progress, transitioning from its α- into its r-phase, it is its core principles that require testing and I begin that process in the following pages. Building on the discussion in the previous chapter, the focus remains on the governance of geoengineering research, leaving open the fundamentally different questions about how, when and where geoengineering might be deployed at scale in the distant future. But first some thoughts on governance generally.

Managing the downside

Governance

There is a vast literature on governance, including now on the governance of geoengineering. Much of this typically follows the tried and tested top down formula. For transboundary issues such as geoengineering, the favoured regimes take the form of a global institution, often an agency of the UN (House of Commons 2010; Shepherd *et al.* 2009). But if this book has one message, it is that a global controller, and even more so, a global global controller, greatly increases the likelihood that an accident waiting to happen, will actually happen. Here I revisit governance first by examining our language – the etymology of our key words conceals considerable ancient wisdom.

The word governance comes from the Greek and Latin words meaning to steer as for a seagoing vessel. Steering has two purposes, first to follow a course that will take you to your destination, and second to do so safely. The notion of steering implies directed motion. Only systems in motion, systems in flux, require governance. If the objective is to prevent action then no governance is necessary and regulation will suffice.

The demand for safety may require some deviation from the most direct, and potentially quickest, route because it is usually better to arrive late than not arrive at all. The deviations from the most direct route are analogous to the redundancy that increases the resilience of a complex system. These deviations will be kept to a minimum because while the steersman wants to arrive safely he also wants to arrive as quickly as he sensibly can using the minimum of resources. Efficiency is important but is subordinate to effectiveness. An effective system of governance will balance these competing demands.

Destination

The Latin roots of *destination* combine a sense of completeness (from the prefix *de-* as for example in *denude* meaning to strip completely bare) and the notion of standing. The word is imbued with a sense of a fixed, unchanging place in contrast to the continuous change as one progresses along the journey towards it. It is perhaps here that the notion of governance is challenged by geoengineering. Where is this fixed place on which we shall firmly stand once we have geoengineered the planet? Consider the metaphor of a race. The destination is typically a line across the track, well defined in space. It is further well defined temporally by the number of circuits the competitors must run or ride. They need to know when to stop. But for climate change there is no track marking a predefined trajectory into the future. There is no line marking the end point. There is no limit to the number of circuits. There is no stopping point. The concept of a destination does not accord with the dynamics of climate change.

The destination for the climate journey is more akin to that of the great explorers of the past, striking out into uncharted, unknown territory. Their destination was not a predetermined place to which they could take the most direct route, always allowing for the necessity of staying safe. Their journeys were more in the nature of meanderings; they had a rough idea of where they were headed but they didn't know their destination until they arrived. In many cases they arrived at places quite different from those they initially set out to discover. When Columbus discovered the Americas he famously thought he was en route to Asia. The future is the temporal equivalent of this *terra incognita*. It is a place to which we travel but can never arrive because it constantly recedes as we approach, much like Tantalus, unable to grasp the fruit that always remained just out of reach. As a destination the future is little more than a tease (is it this or is it that?), a mirage (is it what it seems?), perhaps even a Siren tempting us to foolhardiness (the opposite of what it seems). The notion of an explicit target that defines the endpoint of the geoengineering adventure is misplaced. Rather there is a fog into which we launch ourselves and our prime concern is not to founder on any unseen rocks.

Regulation

Regulation and governance are similar but not the same. *Regulation* comes from the Latin *regula* meaning a ruler or straight edge. This in turn uses the *reg-* root that denotes a king or leader from the Sanskrit *raj*. Regulation connotes some kind of top down imposition of behaviour. The ruler directs – both as king and as straight edge. For a leader to establish rules for others to follow she must have some prior notion of precisely how she wants the activity to unfold. Regulation is generally more prescriptive than simply setting down broad principles such as fairness and efficiency, or exhortations to comply with established norms. Regulations set out in detail how those principles are to be respected in practice by imposing explicit limits on behaviour. These limits may be prudent where the consequences of disregarding them are known to be negative; they may be imprudent when they unduly restrict personal freedom. They correspond to the connectedness dimension of a complex adaptive system and as already explained, while some connectedness enhances the building of system capital, a surfeit reduces resilience making it more vulnerable to surprise misfortunes. While regulation will certainly have a role to play in the governance of geoengineering, care needs to be taken to ensure that it does not become the tail wagging the governance dog.

The role of governance

If the task of delivering us to a destination can be dispensed with, the role of governance is reduced to one of securing safety. In order to protect the diversity that enhances opportunities for innovation, geoengineering governance need not be concerned with what research is done, only that whatever is done, is done without taking undue risk. The great explorers took care, not always successfully, to avoid the many perils awaiting them. Our exploration of the future will continue and at every stage we must endeavour to prevent it coming to an abrupt end by hitting uncharted rocks, or a slow end by gradually decreasing the carrying capacity of our ship. But safety does not mean avoiding risk at all costs. Planet Earth is a ship in unstoppable motion, there is no safe harbour in which to see out a storm. It is for governance to ensure that whatever turbulence arises, it can be accommodated, not necessarily without some local losses, even local catastrophic losses, but certainly without globally catastrophic losses. Unlike the explorers of the past whose catastrophes were local, sometimes resulting in aborted missions, other times in lost lives, geoengineers are exploring processes that have systemic implications if deployed at scale. This entails much greater risk and correspondingly much greater responsibility.

On this reading, the task of geoengineering governance is clear. It is to establish how geoengineering might be used alongside other actions to reduce the overall danger from climate change. If circumstances demand that we steer between the climatic equivalents of Scylla and Charybdis, then we must, but it would be wise to reduce the risks to the extent we can and if there is a possibility that a little geoengineering prudently deployed might help then it would seem reckless to reject it by failing to do the necessary research.

A simple analogy from financial investment illustrates the point. Investors are always warned that values can go down as well as up. In order to protect against catastrophic loss the advice is to diversify, to hold many stocks or even invest passively in the market index. Doing this automatically reduces the maximum gain because any amount not invested in the investment that increases most in value will result in a sub-optimal outcome. But since at the outset, the investor doesn't know which investment that will be, diversifying the portfolio has the effect of reducing the impact of both losses and gains. This risk hedging strategy is designed so that the investor can remain invested and reap some of the rewards from the risks she takes, albeit by sacrificing any possibility of attaining the absolute maximum. This requires that the risk of major losses be rendered insignificant while the capacity to absorb smaller losses remains intact. So it is with geoengineering. The geoengineering proposition is that a judicious mix of one or more SRM or CDR technologies, at a climatically significant but not dominant scale, together with as much emissions abatement we can achieve and as much adaptation we can build, could result in a lowering of the overall climate change risk but not its elimination. But in order to establish that judicious mix, small scale research is necessary and the governance that ensures the safety of that research needs to be commensurate with the scale of the activity.

Despite the occasional local disaster, the established fact is, that where proper governance is in place, those charged with constructing our built environment have shown themselves to be responsible citizens with considerable experience and proven skill at managing risk. The initial challenge of the governance of geoengineering research is to ensure that it is undertaken with that same traditional prudence. I have proposed the Geoengineering Governance Network as the first incarnation of such a system and it proceeds from the assumption that our climate scientists and engineers are a force for good. Research governance is needed to help and support them in devising ways of managing the threats of climate change, not to undermine them in their genuine efforts to make the world a better and safer place.

Subsidiarity

To avoid the systemic threats from a governance or regulation global controller, the principle of subsidiarity is helpful. Subsidiarity requires that the central authority perform only those tasks that cannot be performed effectively at a more immediate or local level; yet again, effectiveness trumps efficiency. The heterogeneity that is vital to the vibrancy of all complex adaptive systems, is strengthened by deliberately having multiple decision makers precisely in order to increase the opportunities for governors and regulators themselves to benefit from the emergence of new ideas and practices that will retain their continuing relevance and effectiveness in the face of the continuously changing present. In this way the risk inherent in a global global controller is diluted by having many local global controllers. The danger that local controllers might act imprudently, whether deliberately or not, is not eliminated but is greatly reduced by the requirement of transparency and continuous scrutiny.

These considerations apply equally to geoengineering research particularly because by its very nature it may be assumed that it is an activity that some in the wider community would want to engage with, even more so should large scale experiments or deployments begin to look more likely. As for exposure to malevolent or undisciplined forces, as explained shortly, these are most unlikely to pose a threat that cannot be contained before systemic damage occurs.

The baby and the bathwater

The lack of significant empirical knowledge about almost every geoengineering technology creates a space in which conjecture rules. Two studies by Bellamy and others (Bellamy *et al.* 2013; Bellamy, Chilvers, and Vaughan 2014) illustrate how speculation, some better and some less well informed, fill the vacuum created by a lack of empirical evidence. The rankings of geoengineering options that they produce from their interviews with both experts and citizens are not based on empirical evidence and must therefore be understood as subjective perceptions rather than as objective assessments of effectiveness and feasibility. While it is valuable to acknowledge these perceptions, care must be taken not to imbue them with the solidity of facts upon which to base decisions about which geoengineering proposals to pursue further. The danger of being swayed by factoids is ever present in a discourse in which there is so little real empirical evidence.

The balance of risk and benefit from any given geoengineering technology, or combination of them, will not become apparent until empirical studies are undertaken that convert incommensurable uncertainties into commensurable risks that can then be used to inform policy. However, even armed with that knowledge, where the balance between too much risk and too little benefit from geoengineering is to be drawn will depend also upon perceptions of the risks from *not* geoengineering, and these are likely to undergo considerable change as the anticipated ravages of global warming begin to impact more noticeably on the daily lives of the world's urban middle classes. If, based on current speculation, geoengineering options are prematurely closed down, those confronted by those more pressing climate threats in the future might rue the excessive caution of their predecessors.

Geoengineering Governance Network (GGN)

The GGN is a modest proposal to establish a research governance regime that draws together the lessons of complexity theory with its emphasis on diversity and emergence (Levin 2003), the insights of cultural theory and its recognition of the need for 'clumsy solutions' (Verweij *et al.* 2006) that recognise the imperative of accommodating the actors' conflicting worldviews, even if imperfectly. It also accommodates the need to escape the constraints of conventional problem solving identified in Cox's critical theory (Cox 1981) whenever we are confronted by wicked problems that defy solution (Rittel and Webber 1973), particularly where they turn on postnormal science characterised by uncertain facts, disputed values, high stakes and a need for urgent action (Funtowicz and Ravetz 2003). The GGN is itself a complex adaptive system that will evolve reflexively in response to events.

The GGN is conceived as a virtual network of a global extended peer community whose purpose is to provide geoengineering researchers with the widest possible reaction to their research proposals in order that they have the opportunity to amend them to accommodate any concerns raised. The forum would be for information exchange only, decisions about research protocols would remain with existing decision makers within the research institutions and those, most often public bodies, regulating research and providing funding. The researchers would be expected to publish all research results on the GGN. The extended peer community would be open to anyone interested in participating and their participation would be regulated so as to ensure an appropriate level of propriety and accountability. This brief outline will be developed as the concept is further explored.

An immediate question that arises is how those undertaking research would know whether their project fell within the aegis of the GGN – what distinguishes geoengineering research from not-geoengineering research? In reality, this is not a binary choice because research that does not initially refer to geoengineering may later be adopted by a geoengineering researcher, and research that is rooted in geoengineering may prove to be of greater value elsewhere. Research is not always a tidy linear affair. Leaving aside secret military research that is not in the public domain, once a research project becomes public it will be open to the wider peer community to argue that it should have been referred to the GGN. The researchers involved will be on notice that any further research should be so referred and a failure to do so will have those consequences that emerge as appropriate in such cases. Given that the GGN is at heart an honour system, if it is adopted by the wider community, researchers will be motivated to be more rather than less inclusive in order to protect their academic reputations. Another consequence of this approach is that the GGN community may itself develop criteria by which to triage research projects according to an emerging geoengineering typology. It is precisely this reflexive behaviour that adds value to the community as a whole.

There are several scenarios that are worthy of examination in order to establish how systemic risk would be constrained by such a diffuse and nonbinding system of research governance. They can be consolidated into three high risk categories according to whether the researcher is selfish, selfless or an established professional. These could be characterised as the maverick Geoengineering Greenfinger (Victor 2008), drawing his inspiration from a James Bond villain trying to geoengineer the world for his own nefarious ends, the Geoengineering Ripper, drawing inspiration from Gen. Jack D. Ripper (Diethelm and McKee 2009), the military loose cannon trying to save the world from itself in the movie *Dr. Strangelove*, and the well-meaning boffin who unwittingly promotes a high risk project. Invented for this purpose, I shall dub her Professor Erica Kilthead. She is a serious academician of considerable integrity and standing. She has advised many national and supra-national agencies on a range of climate change related issues including geoengineering. She is also an active researcher and is able to access financial and human resources to undertake her research projects. How would this diffuse unstructured system of governance operate to constrain the excesses of a Greenfinger, or a Ripper, and avert the unwitting systemic catastrophe precipitated by a well-meaning Kilthead?

Geoengineering Greenfinger

It is easy to dismiss all forms of CDR as amenable to subversion by a Geoengineering Greenfinger. Their scale requirements, delayed climatic and diffuse response and cost would make them unattractive to a Greenfinger. Indeed, someone financing a massive programme of tree planting or carbon sequestration would more likely be hailed as a hero than a villain except by those whose livelihoods and lifestyles are disrupted by his annexation of their land. CDR would not be a Greenfinger's *modus operandi.*

The threat of unilateral deployment of SRM is much discussed in the literature (Horton 2011). More recently, the NAS asserts 'A single nation, or even a very wealthy individual, could have the physical and economic capability to deploy albedo modification with the intention of unilateral action to address climate change in a geographic region' (NAS 2015b). However, this danger seems overstated. No form of SRM is a single application process, even mirrors in space would require continual maintenance, as well as defence if their deployment were contested. For any SRM intervention to have long-term climate-changing potential it needs to be undertaken at considerable scale over many decades (NAS 2015a; NAS 2015b). While a Greenfinger may be wealthy enough to finance some SRM activity that might have some local short-lived effects, it is fanciful, I contend, to imagine that any non-state Greenfinger would be able to sustain such an activity for long enough and at sufficient scale for it to be a serious systemic threat.

It is also inconceivable that any geoengineering activity at climate-changing scale could be conducted clandestinely for anything other than a very short period given the extent and sophistication of satellite surveillance now in operation. This raises two questions in relation to a state-based Greenfinger. First, how would others in the international community react to such a unilateral act, and second, what would such a challenge mean for geoengineering research? An international relations liberal whose primary motivation is to increase absolute wealth and well-being irrespective of any loss of power that this may entail relative to other nations, might tolerate the activities of a Greenfinger until such time as they became a serious threat to its own interests. On the other hand, international relations realists whose strategic objective is to enhance their nation's power relative to others even when this results in them being worse off in absolute terms, might regard the rival state's interest in geoengineering as a potential threat to its relative power that would require either that the maverick state's geoengineering activities be stopped, if necessary by force, or that competing activities should be undertaken intended to counteract the effects of the maverick's potential geoengineering, in effect leading to a geoengineering arms race with uncertain long-term political and climatic consequences.

Moreover, if a powerful state were motivated to embark on a substantial long-term programme of geoengineering research and eventual deployment, it must be assumed that the geoengineering would merely be a device designed to achieve some more fundamental national policy objectives that may even have little to do with responding to climate change. If this were done in the face of opposition from the rest of the international community, it must also be assumed that the maverick nation would be prepared to pursue other aggressive and contested policies consistent with those objectives; geoengineering would hardly be singled out as the sole means of achieving such dominant strategic aims. Either nations respect the autonomy of other members of the international community or they don't and in the latter case, if it were not geoengineering it would be something else.

A moratorium or even outright ban on geoengineering research in most countries would not prevent a maverick state from pursuing its own agenda. Indeed, even for realists there is advantage in undertaking geoengineering research within a cosmopolitan structure because active research by a broad community of nations ensures the transparency necessary to assure the realist that his nation's relative power is either not threatened, or if threats do emerge, that they can be addressed at an early stage.

It has been suggested to me that major oil companies or a small group of oil producing countries might embark upon a unilateral deployment of stratospheric aerosol injection by surreptitiously masking it with a natural volcanic eruption. The idea is that the perpetrator(s) would stockpile the necessary materials and have the delivery mechanisms on standby waiting for a major natural eruption. They would then immediately begin the clandestine release of aerosols into the stratosphere so that those monitoring the eruption were led to believe that all the incremental aerosols came from the eruption. As the natural effects of the eruption subside, the added aerosols would be continued in order to maintain them at a continuing higher level than prior to the eruption. It is easy to dismiss this as entirely ridiculous but then who, even thirty years ago, would have imagined widespread suicide bombings becoming a weapon of choice in anyone's struggle to promote their favoured ideology. However, the global scale of SRM and the impossibility of doing it clandestinely for any extended period, require even this scenario to be regarded as at best on the extreme margins of plausibility.

But the threat from a Geoengineering Greenfinger, whether state sponsored or not, would not suddenly appear from nowhere. Even for a maverick, researching geoengineering would be an incremental process, no one will leap from the laboratory and computer model to deploying geoengineering at scale because, if done on their own territory this would entail considerable risks for their own population and immediate neighbours, and if undertaken in international zones, even in the absence of any more robust response from the international community, it would be disrupted by agencies such as Greenpeace. It is difficult to predict how these international political dimensions would unfold but because the long development time would provide a substantial window for diplomatic and other interventions if deemed necessary, there would be ample opportunity for the necessary controls to emerge. Whether this would be a development of the GGN or an entirely new entity, can be left to those in the future confronted by the immediacy of the international political implications of deployment. However, there can be little doubt that, armed with a comprehensive quality research base, they will be better able to discharge that responsibility, both to the then current generation and their future generations, than without.

In short, the Greenfinger will be constrained by the futility and likely negative consequences for himself, of a long-term contested attempt to deploy SRM geoengineering at scale. Horton (2011) concurs, also arguing convincingly, at least in regard to SRM through stratospheric aerosol injection, that there would be no incentive for such unilateral state action. No form of governance, short of brute force, would have any effect on a truly malevolent lone actor but, other than brute force, none would be needed because if the Greenfinger were to submit to a formal governance process he would transition into one of the other risk categories.

Geoengineering Ripper

Ripper is a serious threat because he personifies a member of the elite with considerable power who has lost faith with due process and is so convinced that he is right that he abuses his power, exceeding his authority in order to force through his own (probably) crazed agenda. In the context of the fictional Dr. Strangelove such a character is a useful device to illustrate the follies of power. However, in the real world an extraordinary amount of both complicity and duplicity would be needed for a Geoengineering Ripper to be able to wreak havoc. Any geoengineering project with climate-changing potential would involve many actors, including individuals, corporations and even government agencies. The comments already made about satellite surveillance apply equally to a Geoengineering Ripper; he might be able to do it once or even twice but not on the continuous basis necessary for it to be a real systemic threat.

If the Ripper posts his project on the GGN it would quickly become public and be subject to considerable scrutiny. Voices from across the global community would be raised against it. How these voices would be translated into the power to stop or modify the project would depend upon the specifics of the case and the mood of the moment. If the Ripper didn't post his project on the GGN or did so misleadingly, he becomes a Greenfinger and the earlier comments apply.

In the few geoengineering experiments so far conducted outside the aegis of formal academic research, all have brought about a closing of ranks rather than any enthusiasm to replicate these privately motivated experiments (Bracmort, Lattanzio, and Barbour 2010, n. 11 and 12). This reflexivity suggests that such a diffuse system of governance would not only be effective at controlling the excesses of potential Geoengineering Rippers, but also be able to do so in a way that continuously adapts as the future emerges, reflecting both changing social priorities and technological advances.

The critical features of governance that ensure this outcome are the oxygen of publicity that a transparent governance network provides, and the broad social engagement in the geoengineering community. It should also be borne in mind that unlike Greenfinger, Ripper is working ostensibly within the system, he is a member of the elite. His ability to go off the rails is in large measure due to the incompetence of his peers and superiors and timidity of his subordinates. Conspiracy theorists and those brought up on a diet of *House of Cards*, *Yes Minister*, *The Thick of It* and other fictional and comic representations of power, might argue that such incompetence and duplicity are all too common and it would be foolhardy not to account for them in the design of a governance regime. The danger with such an argument is that it invites a regressive collapse into anarchy because, ultimately, no one can be trusted to be the guardians of the guardians. In fact, the evidence of humanity's rise over millennia suggests strongly that even where such failures have occurred they are quickly absorbed. A more positive view of humanity, and I suggest more realistic, is that at least one of those within the elite who has relations with our putative Geoengineering Ripper, would be in a position to enjoin others to contain him by either diversion or disempowerment. Arguments that this is a naïve reading of the potential of power to corrupt must be tempered by the realities of the physics, chemistry and engineering of all forms of geoengineering that render them impossible to undertake clandestinely at a scale and over a period long enough to cause irreversible and extensive damage.

Professor Kilthead

Kilthead is a person of integrity who operates within the system and willingly follows its established protocols. Our concern is that she unwittingly precipitates a geoengineering catastrophe. In the first instance we can assume that the threat of a catastrophe has not been identified by the normal routines of research oversight that will have been properly conducted in Kilthead's research institution. We may also assume that Kilthead is a prime mover behind the GGN and has posted her project proposals for scrutiny by the extended peer community. The potential for a catastrophic outcome from the project would also have to have escaped this wider audience. It is plausible that even given this amount of combined examination, the risk of a catastrophic outcome could be missed but it would also be consistent to assume that in this event it would be most unlikely, although not necessarily impossible. Were this not the case, it would either require many people to have failed to act on available knowledge or for the risk to reside in something as yet unknown. The management of risks from the realm of surprise can only be achieved through building safe-fail systems to enhance resilience. No system of governance, including moratoria and outright bans, can be entirely secure against the unexpected. But surprises can be reduced by ensuring the widest possible scrutiny of any proposals because what may be a surprise for some may not be for others. Since geoengineering is likely to remain a highly contested subject, the level of scrutiny offered by an open geoengineering governance network is likely to be the most secure way of highlighting a potential catastrophe from Kilthead's proposal in good time for her to modify it appropriately.

The underlying control mechanism that would be deployed to as near as possible eliminate the risk of catastrophic failure of a geoengineering experiment, is a limit on scale. If the radiative forcing, duration and spatial extension of Kilthead's project exceed what is at the time considered to be a safe threshold for the specific type of experiment she proposes, we may assume that because she is a person of academic and personal integrity, she will modify her project to accommodate the feedback from the GGN. On the other hand, because the governance regime is not mandatory, she might decide to ignore or reject the feedback and proceed as originally planned. In this event she might be obliged by her institution's internal risk management protocols to modify her project. But if she is able to convince her local controllers that the risks have been exaggerated, she may still proceed as planned. Such a move would either be a piece of inspired research that moves us radically forward either to promote or demote geoengineering as a part of the climate change policy mix, or would transform Kilthead into a Ripper or Greenfinger, in which case the earlier comments apply.

It is not hard to imagine that if it were properly administered, the GGN could quickly acquire sufficient status to be an effective governance regime for all reputable institutions and researchers, and the basis for possible future governance regimes that become more institutionalised as the implications of geoengineering become clearer, particularly if research projects become more ambitious and approach deployment scale. Indeed, it may become the case that any researcher who does not engage with the GGN in an open and comprehensive manner will quickly become academically marginalised. This would be a powerful incentive to comply.

Objections to the Geoengineering Governance Network (GGN)

Conceptually, a key concern about the GGN would be that it might become an anarchic forum that would be unable to provide clear guidance on any project. This objection could be addressed by appropriate management, including for example, by moderating the network, by having different classes of participants that were administered in such a way as to deter abusive and unconstructive contributions, and by requiring all participants to be properly identified in order to establish the provenance of each contribution. The GGN needs to provide helpful and considered feedback rather than decisions, and certainly it needs to suppress *ad hominem* arguments to ensure that responses focus on the issues. Decisions would continue to be made by those controlling the projects and in particular those controlling the funding, which is likely to continue the involvement of public institutions in one form or another.

A further objection is that the GGN could be suborned by the geoengineering elite as a means of promoting geoengineering even in the face of significant objections from elsewhere. This objection stems from a misunderstanding of the role of the GGN. Its purpose is initially restricted to that of a risk management tool in the design of research protocols. By giving the maximum exposure to proposed projects and the results of past projects, there is the possibility of much wider engagement across all segments of the global community to question and raise concerns. This would be a reification of Beck's 'unbinding of politics' (1992) and Funtowicz and Ravetz's 'extended peer community' (2003) precisely for their intended purpose of reducing risk in the face of intractable uncertainty. Contributors would have no direct power to approve or veto any project, but on the basis that comprehensive transparency is to be observed at all stages, there would be considerable pressure on those promoting and financing research projects to ensure that the concerns raised in the GGN had been responsibly and comprehensively addressed prior to implementation. Moderators could ensure that only contributions that served this purpose were posted.

Some may argue that the GGN's inability to enforce any particular standard of conduct would allow an effective free-for-all. However, this type of non-binding pressure is a common feature of many long standing and effective international agreements sometimes referred to as 'soft law' (Hillgenberg 1999). Prominent examples of such international soft law include most resolutions and declarations from the UN General Assembly. Another is the Universal Postal Union (UPU) that regulates international post. Membership of the UPU is voluntary but now includes all 192 nations of the world. Hillgenberg cites many reasons for states preferring soft law agreements, several of which are directly applicable to the case of geoengineering. These include:

- a general need for mutual confidence-building;
- the need to stimulate developments still in progress;

- the creation of a preliminary, flexible regime possibly providing for its development in stages;
- impetus for coordinated national legislation;
- concern that international relations will be overburdened by a 'hard' treaty, with the risk of failure and a deterioration in relations;
- simpler procedures, thereby facilitating more rapid finalisation (e.g. consensus rather than a treaty conference);
- avoidance of cumbersome domestic approval procedures in case of amendments;
- agreements can be made with parties that do not have the power to conclude treaties under international law;
- agreements can be made with parties that other parties to the agreement are not willing to recognise.

A significant benefit of non-binding governance structures is that they are more flexible and responsive to circumstances as they unfold. The lapsed time required to develop or amend any mandatory international treaty is typically measured in years. For example, even after 25 years the UNFCCC has yet to produce any effective agreement on emissions abatement, and despite mounting pressure to achieve this at the 2015 COP21 in Paris, many believe that this remains an elusive goal. Although this is an extreme example, even treaty changes under a relatively non-contentious subject, such as dumping at sea, which is regulated by the London Convention and Protocol, typically take years rather than months to implement. Any geoengineering governance regime dependent upon such an extended process would quickly fall into disrepute and be marginalised.

It is conceivable that those ideologically opposed to geoengineering might conspire to overwhelm the GGN by spamming it with responses to a geoengineering proposal. Clearly if very large numbers of responses were received it could become operationally difficult for them to be summarised into useful inputs by those presenting geoengineering research proposals, and also might have the effect of reducing the impact of potentially valuable contributions. This danger might be addressed by developing automated processes for screening, categorising and condensing responses posted to the GGN. It would also be reduced by having robust registration procedures to ensure that all contributors were identified and traceable as well as procedures that discarded abusive postings.

A further objection to the GGN is that it does little to address the slippery slope, moral hazard and several other ethically driven arguments against geoengineering. The difficulty presented by all these arguments is that their proponents use uncertainty to foment fear of the unknown and, using the precautionary principle, conclude that the activity is too risky to undertake. The slippery slope argument maintains that one should not research geoengineering now, despite the possibility that it may be wise to do so, for fear of raising expectations that it will lead to full deployment in the future, a prospect that is currently considered undesirable. This reticence denies the possibility that those in the future armed with greater knowledge about both the risks and benefits of geoengineering might take a different view. The difficulty with this argument is that its proponents cannot prove that deployment of some geoengineering as part of an integrated response to climate change will be undesirable because if they could that would be a sufficient reason not to do it and the slippery slope argument would be superfluous (Cornford 1908; Keith 2000).

The moral hazard argument is that resources should not be devoted to geoengineering because that would reduce those available to more worthy responses to climate change. This fails on the same principle as the slippery slope argument because its proponents can show neither that investment in geoengineering would reduce that available to other responses, nor that some geoengineering in the policy mix would not significantly enhance the overall response to climate change. If they could show that in practice geoengineering is redundant because there is high confidence that other responses are sufficient, the moral hazard argument would also be redundant.

In so far as other ethical arguments reflect the changing social values of the moment or those from diverse social settings, the GGN would, by virtue of its wide public engagement, be a reflexive mechanism for ensuring that those values were considered in assessing any geoengineering proposal. This is precisely the opposite of relying on an elite to make such judgements because the processes whereby the members of the elite are selected greatly increase the danger that their judgements will be partial. The elite may favour an outdated *status quo* or powerful vested interests on whose patronage some of them depend for their own status, or simply fail to consider the views of those whose voices rarely reach the ears of the elite. On balance, there seems no reason in principle why a GGN would not provoke at least as challenging an ethical assessment of any geoengineering proposal as any more centralised governance process. In practice, its effectiveness at doing so would be determined by the number and range of participants and the respect their views are given by the project sponsors. This can only be determined heuristically.

The credibility of the GGN could be undermined if it were perceived to be a creature of the academic geoengineering community or any other interest group. This is not to say that the great majority of senior academics or policy advisors are not well versed in exercising independent judgement, but as and when contentious issues arise as they surely will, if the GGN were too closely identified with them, avoidable complications could arise – not only must justice be done but it must be seen to be done. This difficulty could be averted by having the GGN hosted and moderated by an independent body especially created for the purpose and perhaps sponsored by learned societies such as the UK's Royal Society or the US National Academy of Sciences or some collaboration between them and their counterparts from other countries. Alternatively the GGN would sit well within one of the UN's environmental agencies, perhaps the UN Commission on Sustainable Development or UNEP.

The SRMGI is an early example of elements of such an approach. It was organised by the Royal Society and related organisations from the US and the developing world, and undertook a consultation on the governance of SRM.[1] While it produced some useful insights it was not constructed as a proto-governance system and has not been developed more fully. Establishing credible independence is another practical detail that can emerge as the concept of the GGN progresses through its α-phase. However, care would need to be taken not to allow the management of the GGN to be dominated by climate scientists or geoengineers; its role is limited to being an orderly forum for delivering transparency in geoengineering proposals and results, and as such the skills required for its management lie more within the domain of communications than in climate science or engineering.

Other risk factors associated with geoengineering include the technical control dilemma, reversibility, encapsulation, commercial involvement and lack of public engagement (Bracmort, Lattanzio, and Barbour 2010). The technical control dilemma refers to the idea that the impacts of new technologies are often difficult to discern prior to their full development, yet become increasingly difficult to control once they become entrenched. This has much in common with the slippery slope argument because it expresses not simply that the governance of technology becomes more challenging as it becomes more widespread, but, significantly, the hubristic fear that future generations will not have the collective wit to meet those challenges. I return to this issue in the next chapter.

Reversibility and encapsulation (the avoidance of risk by not introducing novel agents into the environment) are concerns directly related to scale. It does not follow that hazards would always be reversible even if the scale of geoengineering research interventions were properly controlled, they could nevertheless be manageable in accordance with the principles of safe-fail.

Risks associated with commercial involvement require further thought because at present, what little evidence exists about commercial involvement, suggests that there is no appetite to engage in geoengineering research amongst major corporations with the capacity to do so at seriously risk-inducing scale. At present it seems unlikely that commercial entities will engage in major geoengineering research projects other than those sponsored by academic institutions or state agencies. This being so, that research would be governed by the participation of those institutions in the GGN. Even if commercial entities were to become independently engaged in geoengineering research, if the GGN acquired the necessary status, it would be difficult for any commercial activity not to be mediated through it if the companies concerned were not to risk becoming Geoengineering Rippers. For reputational as well as financial reasons, it seems most unlikely that they would incur that risk.

As regards Bracmort *et al.*'s concern about public engagement, few governance systems could be more open to public engagement than the proposed GGN. The critical questions will be about how that is operationalised so as to ensure its credibility and legitimacy with the public. This is not the place to develop these detailed procedures, but these would emerge through experience should the GGN become a reality.

The active debate within the academic geoengineering community about the importance of governance suggests that they would be receptive to participating in the GGN provided they believed that it would be effective in both promoting research and managing risk. There is an inevitable challenge of building the critical mass to deliver this but this is not a knock-down argument against the GGN. One might equally ask why would anyone have bought the first telephone or fax machine when, in their early days, there were so few others with whom to communicate. History suggests that if a new service offers real benefits, it is quickly taken up, and conversely soon expires if it is not perceived as adding value. Without making the attempt, it is not possible to know whether the GGN could become the *de facto* geoengineering research governance regime. Whether this attempt should be made depends on their being more promising alternatives.

GGN – legitimacy and credibility

The critical tests for the GGN will come from three constituencies. First, will those wanting to undertake empirical research into geoengineering recognise the benefits that will accrue to them from the GGN's ability to bind them into a wider social community? Any regime of governance will introduce new administrative demands that some will consider burdensome. One of the marks of success is that the benefits from the governance regime for those being governed outweigh its inconveniences. This can only be established heuristically. For those with a genuine commitment to examine the potential of geoengineering to serve humanity's long-term interests, the more comprehensive, the more transparent, the more vibrant the GGN community, the more it will add value to their own research.

The second constituency is civil society, including those that are opposed in principle to geoengineering and the general public and media who wish to be informed and to have the means of registering their views. They should welcome the GGN because for the first time it will provide them with a comprehensive register of geoengineering research proposals, informed comments on them from across the academic and wider social spectrum and access to the research results. In addition it provides them with a unified platform on which to register their concerns although it in no way interferes with their rights to protest in other ways. On the basis that knowledge is power, the GGN will provide immense power both to those opposed to and those wishing to promote geoengineering. But this power will be constrained. The GGN is a forum for the exchange of information, it is not a decision making entity; decisions will still rest with the researchers, their funders and sponsors. Whether civil society will consider access to such a complete dataset on geoengineering activity to be worth supporting despite it not providing them with a power of veto must remain an open question until it is tried.

The third constituency is the policymakers, comprising the civil servants who advise politicians, and the politicians themselves. The interests of this group are vital because in one way or another they will be responsible for most of the funding for any empirical geoengineering research. This must be so because geoengineering rarely produces any marketable end product, and in those few instances where it does, for example captured CO_2, it does so on such a vast scale as utterly to overwhelm the potential market, or the end products are produced as by-products in insufficiently large quantities and at too high a cost to be commercially significant, for example electricity generation from biochar.[2] If geoengineering is seen as a public good, it will be financed through and governed by the public purse. Assuming that the other two constituencies engage wholeheartedly with it, the significance of the GGN for the policymakers is that it demonstrates the credibility, legitimacy and social acceptability of the governance processes being followed by researchers. For this to happen, the participators in the GGN must be seen to act with respect for the integrity and motivations of each other. The researchers must be truly transparent and comprehensive in their postings on the GGN and those commenting on them must be measured in their reactions. Importantly, the researchers must also explain how they have accommodated (or not) the objections and concerns raised by respondents when finalising their research protocols.

The academic geoengineering community is currently concentrated in a handful of countries, notably the US, UK, and Germany although many others are also active. This is a small community of perhaps less than a thousand individuals. The GGN offers a means for this community to become truly global, sharing information, arguing, learning, and generally engaging in a clumsy inefficient slightly chaotic discourse from which can emerge multiple initiatives that can combine in an adaptive manner and so guide the policymakers to make timely, effective and low risk decisions that serve global long term priorities but respect local short term concerns.

Conclusion

Geoengineering can only be deployed as a credible response to climate change as part of an integrated policy mix in which it does not play a dominant role. Geoengineering must be undertaken at scale and continuously for several decades to have any enduring climate impact and this largely eliminates the risk that it could be used other than within a broad international consensus. Whether geoengineering can in fact enhance humanity's response to climate change by being part of the policy mix will not emerge until empirical research is undertaken, and if it can, how it can will be subject to continual enhancement with the benefit of experience.

The risks from geoengineering research are best controlled through a process of wide non-binding consultation that is transparent as regards both research proposals and results, and independently managed. Such a process would ensure the widest possible engagement, bringing to those promoting geoengineering the views of a culturally diverse range of global actors. This process would be autonomously flexible and responsive to changing social values and technological advances allowing research to proceed without the potentially extended delays of more formalised and bureaucratic governance structures. It would also greatly enhance the prospect that decisions about geoengineering research would be free from domination by any elite group either protecting narrow vested interests or pursuing a self-interested ideological agenda.

Through the GGN, the process of geoengineering research governance becomes a complex adaptive system in its own right, both nested within and operating alongside the multitude of other complex adaptive systems that constitute subsets of the ecosphere. This structure enables and encourages the non-linear thinking required to promote the innovation needed to meet the challenge of climate change. Risk management is also an emergent property of the diversity and reflexivity of this complex adaptive system. As empirical experience builds both theoretical and practical understanding, policymakers will have a range of viable geoengineering options, as those that have proved less apt, whether in terms of cost effectiveness or risk profile, will have been eliminated. The autonomous process that produces this outcome will have permitted ever larger scale experiments of those geoengineering interventions, if any, shown by the empirical evidence to be most effective and least risky so that the policymakers of the future will be able to act on the basis of knowledge rather than on conjecture, ideology, or unproven notions about how the world works. The words 'effective' and 'least risky' are open to a wide range of interpretations. It is the job of policymakers to interpret these concepts in ways that are sensitive to their prevailing political, social and climatic circumstances. These will vary spatially and temporally and the argument from complexity is that while eliminating the patently dangerous and ineffective, policymakers should keep as many of the plausible options open as possible so as to fuel the process of innovation.

As appropriate, the governance regime itself will undergo transformation to accommodate the shift from the political demands and risk profiles of initial proof of concept, to the very different political demands of intentional climate-changing deployment at scale. This may involve the creation of an entirely different governance regime, but the format this may take is a matter for those in the future who, unlike today's decision makers, are confronted by the reality of deployment decisions.

A research focused GGN could be established quickly at minimal cost. Its role would be established *de facto* by the way in which it is embraced by the global geoengineering community of academics of all relevant disciplines from the climate scientists to the philosophers, the media, policymakers, civil society organisations, and interested member of the public. The involvement of this extended community is essential to minimise the impact of groupthink within the academic community. The GGN is a Protean concept that will both shape and be shaped by the future.

Notes

1 Refer to the SRMGI website for further information, at www.srmgi.org/.
2 The tens of thousands of biochar installations needed to sequester $1PgCyr^{-1}$ (see Appendix to Chapter 2) could produce meaningful amounts of electricity that could have beneficial local impacts. However, it is far from clear that energy companies would invest in this technology as a commercial enterprise because the additional costs of sourcing the biomass feedstock and disposing of the biochar and the waste products would make these installations high cost producers of electricity. As a result, biochar would only be an attractive proposition for the power industry if these additional costs were covered by public subsidies.

References

Beck, Ulrich. 1992. *Risk Society: Towards a New Modernity*. Sage Publications Ltd.
Bellamy, Rob, Jason Chilvers, and Naomi E. Vaughan. 2014. 'Deliberative Mapping of Options for Tackling Climate Change: Citizens and Specialists "Open Up"appraisal of Geoengineering'. *Public Understanding of Science*. doi:10.11770963662514548628.
Bellamy, Rob, Jason Chilvers, Naomi E. Vaughan, and Timothy M. Lenton. 2013. '"Opening Up" Geoengineering Appraisal: Multi-Criteria Mapping of Options for Tackling Climate Change'. *Global Environmental Change* 23(5): 926–937.
Bracmort, Kelsi, Richard K. Lattanzio, and Emily C. Barbour. 2010. 'Geoengineering: Governance and Technology Policy'. In Congressional Research Service, Library of Congress.
Cornford, F. M. 1908. *Microcosmographia Academica*. Bowes & Bowes Ltd. www.maths.ed.ac.uk/~aar/baked/micro.pdf.
Cox, Robert W. 1981. 'Social Forces, States and World Orders: Beyond International Relations Theory'. *Millennium – Journal of International Studies* 10(2): 126–155. doi:10.1177/03058298810100020501.
Diethelm, Pascal, and Martin McKee. 2009. 'Denialism: What Is It and How Should Scientists Respond?' *The European Journal of Public Health* 19 (1): 2–4.
Funtowicz, Silvio, and Jerome R. Ravetz. 2003. 'Post-Normal Science'. International Society for Ecological Economics (ed.). *Online Encyclopedia of Ecological Economics*. www.Ecoeco.Org/publica/encyc.Htm.
Hillgenberg, Hartmut. 1999. 'A Fresh Look at Soft Law'. *European Journal of International Law* 10(3): 499–515.
Horton, Joshua. 2011. 'Geoengineering and the Myth of Unilateralism: Pressures and Prospects for International Cooperation'. *Stanford J Law Sci Policy* 4: 56–69.
House of Commons. 2010. 'The Regulation of Geoengineering'. House of Commons Science and Technology Committee. www.publications.parliament.uk/pa/cm200910/cmselect/cmsctech/221/221.pdf.
Keith, David W. 2000. 'Geoengineering'. *Encyclopedia of Global Change*. Oxford University Press.

Levin, Simon A. 2003. 'Complex Adaptive Systems: Exploring the Known, the Unknown and the Unknowable'. *Bulletin of the American Mathematical Society* 40(1): 3–20.

NAS. 2015a. 'Climate Intervention: Carbon Dioxide Removal and Reliable Sequestration'. National Academy of Sciences. www.nap.edu/catalog/18805/climate-intervention-carbon-dioxide-removal-and-reliable-sequestration.

NAS. 2015b. 'Climate Intervention: Reflecting Sunlight to Cool Earth'. National Academy of Sciences. www.nap.edu/catalog/18988/climate-intervention-reflecting-sunlight-to-cool-earth.

Rittel, Horst W. J., and Melvin M. Webber. 1973. 'Dilemmas in a General Theory of Planning'. *Policy Sciences* 4(2): 155–169.

Shepherd, John, Ken Caldeira, P. Cox, J. Haigh, David W. Keith, B. Launder, Georgina Mace, G. MacKerron, J. Pyle, and Steve Rayner. 2009. 'Geoengineering the Climate: Science, Governance and Uncertainty'. The Royal Society.

Verweij, Marco, Mary Douglas, Richard Ellis, Christoph Engel, Frank Hendriks, Susanne Lohmann, Steven Ney, Steve Rayner, and Michael Thompson. 2006. 'Clumsy Solutions for a Complex World: The Case of Climate Change'. *Public Administration* 84(4): 817–843. doi:10.1111/j.1467–9299.2006.00614.x.

Victor, David G. 2008. 'On the Regulation of Geoengineering.' *Oxford Review of Economic Policy* 24 (2): 322–36. doi:10.1093/oxrep/grn018.

9 Drawing the threads together

As I noted in Chapter 1, it is often said that there's nothing new under the sun. Geoengineering is. For the first time since the planet's formation some 13 billion years ago, one species is contemplating intervening at a global scale in the physical processes that determine the surface and atmospheric environment for all species. The advent of the Anthropocene recognises that humans increasingly have the power to shape the planet but hitherto this shaping has arisen from the accumulation over millennia of multiple local interventions, some intentional, others not (Lewis and Maslin 2015). But the intentional intervention at a global scale is novel. Some regard this as a bad idea in principle (Hulme 2014); others regard it as regrettable but potentially necessary to avert the worst ravages of climate change (Keith 2013). I regard it as a challenge for which evolution has not prepared us.

Much of the discourse around climate change is concerned with the assessment of risk and the notion of 'acceptable' risk. Our perception of acceptable risk varies according to what is at stake. We expect much lower risk from aircraft than from refrigerators. The hazard of dying in a plane crash sets the level of acceptable risk much lower than does the hazard of having no ice cream. The question is where on this spectrum do we locate the risk from dangerous climate change. If we set it at or close to the risk of dying in a plane crash then we should be prepared to take the consequences of reducing the climate change risk to the same low level. However, the evidence from the graph in Figure 1.1 is that we place the hazards from dangerous climate change somewhere below the hazard of having no ice cream despite a general awareness that this inverts their real relationship. The reason for this, I suggest, is first that the hazard of having no ice cream affects me here and now, whereas the hazards from dangerous climate change, however much more severe they may be, will not affect me but others in the distant future with whom I have no connection. Second, we unquestioningly assume that there's plenty of time for those in the future to solve the problem. Our evolution from our pre-human ancestors has prepared us well to attend to our own immediate interests but has never needed to promote concerns for distant future generations as an adaptation that enhances the survival of the species. That might be about to become a source of systemic vulnerability.

Over the last decade an increasing amount of theoretical work has been done in order better to understand the range of risks and benefits that might accrue from different technological approaches to engineering the climate but to date almost no empirical studies have taken place outside of a laboratory setting. If geoengineering is to progress from an imaginary to a reality that first step, however small it may be, must be taken. But there is great symbolism in that first step. In cultural theory terms, the battle lines are drawn (Thompson, Ellis, and Wildavsky 1990). The Individualists argue that humanity progresses by exploiting its talents and the resources available to it to address the issues it confronts, while the Egalitarians maintain that we have created the climate problem precisely by the excessive and reckless exploitation of those resources, and prudence and justice now demand that we reduce our planetary footprint rather than make it even more intrusive. The Hierarchists watch from the sidelines ready to mediate if the Individualists prevail, so as to ensure that whatever actions ensue are not too discriminatory or incautious. The Fatalists will endeavour to cope as best they can with whatever awaits them, disenfranchised by their perceived powerlessness to influence events.

For those that consider geoengineering to be a bad idea in principle, discussions about how it might be governed miss the point. For them what is required is not governance of an activity in motion but regulation that stops it dead in its tracks. For the rest, a credible governance regime is required as an essential precursor to any outdoors empirical research. If there is to be any accommodation between these two opposing positions, the Individualists will have to accept some initial constraints that are more restrictive than they would like, and the Egalitarians will have to accept some empirical research provided the risks associated with it are demonstrably low, and everyone will have to rely on both present and future generations not to risk blowing up the planetary laboratory (Charlesworth and Okereke 2010). The Individualists must recognise that progress cannot be bought at any price, and the Egalitarians that the potential for harm does not make the harm inevitable. Indeed, it is possible that averting one harm might provoke another even worse, and on occasions it is prudent to accommodate an earlier harm in order to prevent a later harm being amplified by procrastination and prevarication. By analogy, it is received wisdom in investment management that it is better to take losses early when they are smaller and easier to absorb.

The central messages from a systems view of these challenges are first that there is not, and can never be, one right answer about whether to geoengineer or not, and even less, one that can be determined by any amount of analysis of past events and predictions of future ones. Second, all decisions and indecisions about near-term policy directed at long-term outcomes entail quantifiable risks, unquantifiable uncertainties and inconceivable surprises. While analytical and predictive methodologies alert us to the gravity of the long-term consequences of Business as Usual, it is only through heuristic methodologies that we can step into the unknown future, modifying BAU without compromising the ecosystem's capacity to accommodate and buffer the occasional and inevitable mistakes that we will make along the way.

To be absolutely clear, I am not advocating the deployment of geoengineering, certainly not now, nor even in the future, because we simply do not yet know enough to establish whether geoengineering used in this way would reduce the overall climate risk. But equally, we do not yet know enough to be confident that it would not. The imperative is to conduct research to answer this crucial question. For the reasons explored elsewhere in this book, no amount of theoretical or modelling research relying on predictive methodologies will allow us to transition in one giant leap from geoengineering as a thought experiment to geoengineering as ready-for-action policy tool to advance our cause in the face of climate change. That requires empirical research. Some of that can be conducted at small scale in the laboratory, but in a complex adaptive system the only way to test alternative policy options is to test them on the system itself. There can never be a comprehensive theoretical or laboratory scale analogue for the ecosphere. This being so, experimenting outdoors is necessary to protect the possibility that geoengineering might prove to be a valuable adjunct in combating climate change, or indeed, rejecting it without committing a monumental Type I error.

The argument running through these pages has located the physics and chemistry of global warming within the wider context of the infinity of complex adaptive systems that describe the relations and interactions between all the entities on the planet, as well as the extra-planetary influences upon them. By this framing it explains the intrinsic and inescapable unpredictability of interventionist policies intended to benefit those in the distant future perceived to be at risk from dangerous climate change. It has also argued that this unpredictability must be embraced if policymakers, rather than seeking to reduce uncertainty as a prior condition to action, are to develop risk sensitive near-term policies that encourage the diversity that is necessary for the emergence of the innovations that will be needed more confidently to confront the future. That these innovations are essential is evidenced by the current absence of viable options for decarbonising the global economy at a rate that is likely to prevent the ravages of dangerous climate change from touching ever increasing numbers of people and other biota as this century progresses. The demands for continued global economic growth while simultaneously decarbonising the global economy are far from being reconciled.

Geoengineering is presented as no more than a possible additional component within the policy mix, one that might allow the overall risk from climate change to be reduced, particularly given the increasing awareness of the unprecedented and most probably unattainable objective of averting dangerous climate change by reliance only on emissions abatement and adaptation. Arguments that any form of geoengineering could be an alternative to emissions abatement or an emergency response in the face of an imminent climate catastrophe, are rejected as are critiques of geoengineering that proceed from these assumptions.

The challenge for policymakers is how to embark upon the research necessary to establish whether and how geoengineering might be able to reduce the overall risk from climate change. I have argued that the governance necessary for this process should itself be thought of as a complex adaptive system that will evolve as knowledge and empirical evidence accrue. At the outset the governance should be more focused on encouraging innovation than on determining deployment policies. This balance can shift as the potential benefits from geoengineering become better understood and responses to its risks become increasingly empirically evidenced freeing us from recourse to conjecture and supposition. During this early phase, the possibility must be kept open that there is no form of geoengineering at any meaningful scale that does not entail unacceptable and uncontrollable risks. But crucially, during these early stages that possibility must inform research proposals designed to acquire empirical evidence to substantiate or refute these concerns, and not be deployed as a reason not to undertake this research.

For policymakers to be able to adopt the heuristic approach to policy formulation implied by these considerations, requires a paradigm shift that recognises that all policies intended to address climate change are experimental. By this is meant that they will be designed to test alternative options as a means of edging step by step towards an unknown future, as opposed to attempting to bring into being some pre-imagined future state of affairs. The complexity of the many interrelated systems is such that it is not, and never will be, possible to predict the distant future with sufficient reliability to be confident that pulling any particular policy lever will have any predetermined policy outcome. The only way of moving forward in policy arenas characterised by such complexity is in small incremental steps, constantly monitoring progress to determine whether the direction of travel is towards or away from improvement, and adjusting policies accordingly. However, for climate change this process is made more difficult by the scale and inertia of the climate system. Policy interventions may take decades to become apparent and it is important to resist the temptation to over-prescribe while waiting for a response, only then to discover that the system has been over-stimulated.

Additionally, the paradigm shift must recognise the vital importance of diversity and the dangers from a global controller. In circumstances where there is high uncertainty between intervention and outcome, and action is urgent and values contested, we are denied the opportunity to experiment serially and must encourage many initiatives in parallel. We must also allow for the greatest possible interaction between those initiatives in order to fuel the autonomous process that selects the more apt innovations for enhancement and replication.

Emergence is quintessentially non-linear and while the process is to some degree predictable, the specific emergent outcomes are not. Statistical techniques such as Monte Carlo simulations, designed to cope with randomness, can build reliable probability distribution functions where the empirical data exists to populate the simulations, but they cannot predict which of the outcomes will actually come to pass, merely their likelihood. Possible futures whose probabilities fall in the tails of these distributions, whether thin or fat, are possible and some of them, however unlikely, will happen. Moreover, there are always the surprises, the futures that were not even conceived of before they arrived, or were discounted as being altogether too remote.

Where ignorance is inescapable, the greatest danger is from a powerful elite that is certain that it has 'the answer' and seeks to impose it system-wide. It matters not whether this confidence stems from impartial and objective judgement drawing on the always incomplete empirical evidence available at any moment in time, or on ideology or on the vested interests of the powerful. This danger is greatly reduced by recognising that climate change is not a problem that has a solution but is a situation to be managed and that that management will need to be constantly modified to respond to the unpredictable future as it unfolds. The distinction between certainty and truth is ignored at our peril.

I have proposed the Geoengineering Governance Network (GGN) as the first incarnation of a governance regime for geoengineering that accommodates the needs of an evolving complex process. It is focused only on the early stages of empirical research and I fully expect that it would undergo continual development as the knowledge base grows. It is intended as no more than a starting position to allow empirical research to begin in a manner that is both sensitive to risk and offers the maximum openness to, and opportunity for contribution from, the wider community without stifling the diversity and creativity essential in the research process.

Usurping the future

In these closing paragraphs I feel it necessary to return to the theme of the political rhetoric discussed in Chapter 1, namely the relations between present and future generations. An argument often raised against geoengineering is that it represents a slippery slope to an undesirable future. I have examined and dismissed this argument in Chapter 8 but I revisit it here because it is also an argument that says something powerful about our relations with future generations. The slippery slope argument assumes that we know better than future generations what is best for them, and crucially, makes it a moral imperative not to act in a way that might encourage a future generation to jeopardise its own interests or those of an even more distant future generation. The slippery slope is a transgenerational means by which bad outcomes in the distant future are held to be the moral responsibility of the present. Is this either fair or sensible?

The temporal and spatial reach of humanity's modern technologies creates the need for a new ethics (Jonas 1973). The 'ethical universe', Jonas argues, is no longer composed of contemporaries or confined to those places in which humans physically interact, nor even to humans alone given that 'the biosphere as a whole and in its parts [is] now subject to our power'. He identifies an ethical dilemma. On the one hand our 'knowledge must be commensurate with the causal scale of our action' but on the other it cannot be because our 'predictive knowledge falls behind the technical knowledge which nourishes our power to act'. He concludes that this conflict requires us to recognise that:

> ignorance becomes the obverse of the duty to know and thus part of the ethics which must govern the ever more necessary self-policing of our out-sized might.
>
> (ibid.)

In turn, this demands a new humility based on an awareness of 'the excess of our power to act over our power to foresee and our power to evaluate and to judge'. Modern technologies, and geoengineering in particular, have made the management of this excess of power a central concern for all those affected, and that encompasses all of us. In the early 1970s Jonas was opening up this debate and declared himself 'stuck' because the 'apocalyptic pregnancy of our actions' makes the scientific knowledge that we lack more urgent than ever, yet that urgency is no guarantee of success.

Almost half a century later, the prospect of turning the imaginary of geoengineering into a reality makes the ethical dilemma identified by Jonas a matter of practical concern. Recognising the risks associated with geoengineering provokes us to ask whether we should abandon it altogether, as was done for nuclear powered aircraft and rockets because it was thought that whatever their potential benefits, they could never be justified by the potential hazards,[1] or do we believe that for some forms of geoengineering it might be premature to reach such a definitive conclusion and to do so could be to the serious detriment of future generations?

The temporal horizon referred to by Jonas applies not only to the direct effects of modern technologies but also to their future development. Geoengineering is not a suite of technologies that the present generation will design, test, develop and then mothball awaiting deployment by some indeterminate future generation. Geoengineering technologies are presently little more than concepts and precisely because the potential hazards are so great, we can expect that their development will be slow deliberately in order to ensure that undue risk is not incurred at any stage. However, that its potential risks are so great is a reason to manage those risks, not a reason to squander the opportunity of securing its potential benefits. Moreover, those risks and benefits must always be reckoned relative to the alternatives. Enough is known to suggest that a 4°C warmer world would be a very high-risk environment for many people and many other forms of life.

Those working with these concepts today cannot know with certainty how they might develop over the coming decades nor what criteria might be used by future generations to decide when and how to deploy them. But equally, can they not assume that their successors will be just as concerned about the management of risk as they are? For all its awesome potential, geoengineering is conceived as a public good, not just instrumentally to serve the utilitarian purposes of the living, but more to attend to the interests of future members of the human species as well as the biosphere as a whole. The slippery slope argument requires us to believe that no form of geoengineering could ever reduce the overall risks from global warming and that future generations will act recklessly with the excess power arising from their inheritance of geoengineering.

It is perfectly reasonable to reject these assumptions but to do so has significant implications. It seems wholly unreasonable to suppose that if future generations were to act recklessly with their excess power that they would do so uniquely with that derived from geoengineering. We already have nuclear weapons and there seems every reason to suppose that other sophisticated means of killing people, either en masse or individually, will be developed. As globalisation proceeds apace, all manner of contingent threats are likely to arise from the combined effects of the increased connectedness that this entrains and the limiting of heterogeneity caused by the burgeoning hegemonic power of global elites. These effects would inevitably reduce resilience creating more 'accidents waiting to happen'. Are we to believe that no generation will ever lay the foundations for a technology that some future generation might not abuse? Indeed, how can we know what use a future generation might make of any technology?

Consider stones and intercontinental ballistic missiles. The trajectory from the slingshot through the arrow, the cannonball, and the bullet to the modern missile is largely one of scale – how big and destructive can we make an object and then how far can we accurately propel it? Does this make prehistoric (wo)man in some way complicit in nuclear warfare? The slippery slope argument would suggest so. The Old Testament makes a relevant distinction. When the sins of the father concern his covenant with God, those sins shall be visited upon his successors (Exodus 34:7) but when they are a matter of law, each is responsible for his own (Ezekiel 18:20). This begs the question as to when developing and deploying new technologies amount to sins, and whether, when they do, those sins offend some moral duty or are merely a matter of jurisprudence. These are not questions with straightforward answers. Slingshots, arrows, and bullets all have many benign uses. Armaments generally could be defensive as well as offensive. In other technologies, abortion may be about female emancipation or gender-driven foeticide; a hammer may be a carpenter's tool or an instrument of torture. The Internet and mobile phones are wondrous inventions of the modern age upon which daily life for much of the global population now depends, yet they are routinely used by criminals and terrorists in furthering their nefarious acts. Protecting indeterminate future generations from their own folly seems an extraordinarily hubristic and forlorn enterprise.

These examples suggest that broadly, to the extent there is anything morally reprehensible about a technology it resides in the intentions and actions of the user not in the technology itself. No inventor can be in control of all the uses to which his invention is put or be able to foresee all its consequences into the distant future. Moreover, no invention is independent of its enabling technologies; should any guilt that attaches to an inventor equally apply to all those responsible for the precursors to his invention? Surely this is too extensive; drilling down through the enabling technologies would encompass almost every advance of human society, even writing and language. The ethical issues here are complex and my purpose is to do no more than argue for the unsustainability of arguments that turn on suppositions about the motivation and behaviour of distant future generations (and even the more proximate for that matter – what parent hasn't wondered why his or her child chose that person as a friend, or listens to that music, or has their hair coiffed in that way, or wears those clothes, or had that tattoo or those piercings, and so on). We reproduce intergenerational differences with each new generation.

The choices we make about how to respond to climate change have distributional consequences that will be advantageous to some and disadvantageous to others. This applies not only to geoengineering but also to emissions abatement and adaptation. Almost no social policy of significance is Pareto optimal; policy interventions routinely generate winners and losers and the losers cannot always be fully indemnified for their losses, indeed, in many cases their losses are part of the policy objective – consider redistributive taxes to favour the poor at the expense of the rich, and military personnel called to lay down their lives for the greater good. While debating the intergenerational ethics of geoengineering is of vital importance as we identify and discharge our climate change responsibilities to each other and to future generations, making action contingent upon their resolution results in policy paralysis.

Arming the future

The slippery slope argument assumes that the future generations further down the slope are incapable of resisting temptation to abuse the assets they inherit. This is an extraordinarily arrogant stance for any generation to take relative to its successors. If those in the future decide that emissions reductions and adaptation will not be sufficient to stave off dangerous climate change, as increasingly looks likely, on what moral grounds do we deny them the possibility of using some geoengineering if they judge it prudent to do so?

Stephen Gardiner *et al.* (2010, 284 et seq) call this the 'Arming the Future Argument' and undertake a forceful and insightful examination of it. They conclude that as an argument it 'seems glib, even cavalier' and should be rejected. However, they caveat this rejection in subtle and important ways. Their critique of geoengineering, like many others, is predicated upon geoengineering being a dominant response to climate change. Yet, they explicitly acknowledge that their conclusion might be very different if geoengineering were framed 'as part of some broad climate policy portfolio that includes many [...] alternative policies'. As I have argued throughout this book, geoengineering in whatever form has no scientific justification other than as part of such a portfolio. Although published in 2010, the first draft of Gardiner *et al.*'s book chapter was written in 2007 at a time when geoengineering was much less well researched than it was even in 2010, and when there was much higher confidence than there is today that the UNFCCC would deliver the emissions abatement policies that would render geoengineering redundant. Indeed, at that time, geoengineering was not even considered by the IPCC, receiving only cursory mention in AR4. The shift from AR4 to AR5, bringing some seven years more climate research into consideration, while not completely dashing hopes that emissions abatement might be sufficient to avert dangerous climate change, makes it clear that the hill has become steeper with each passing year of relative inaction. It is necessary to consider whether, with the accretion of knowledge since 2007, Gardiner *et al.*'s caveat needs to be invoked.

The 1,000PgC cumulative emissions target

As discussed in Chapter 1, the focus in IPCC AR5 has turned to cumulative emissions and the current scientific prognostications are that these must be kept below 1,000PgC if GMST increase since the preindustrial era is to be kept below the 2°C that is considered by many to be the threshold beyond which dangerous climate change begins to manifest. But this projection comes with error bars. The 2°C lies in the range 0.8°C to 2.5°C and it is considered to be no more than *likely* that the 1,000PgC of cumulative emissions would result in warming within this range. In IPCC jargon, *likely* means more than a 66 per cent probability – there could be only a two in three chance of the change in temperature falling within this range if the 1,000PgC limit were reached. Furthermore, most of the remaining third lies beyond the top end, there being almost no likelihood of warming being less than 0.8°C, most of which has already happened. If every third airplane crashed, it seems doubtful that aviation would be as popular a form of travel as it is. Accordingly, keeping within the 1,000PgC cumulative emissions should not be considered a low risk option, even more so when it is recognised that the IPCC 66 per cent likelihood only accounts for those risk factors amenable to quantification and these are themselves determined at least in part subjectively based on expert judgement rather than mathematical analysis of observed data (IPCC 2014b, sec. 12.5.4) and we know that experts routinely undercall the risks in their areas of expertise (Kahneman, Slovic, and Tversky 1982). Incommensurable uncertainties and inconceivable surprises are part of our future and if the risks associated with them are accounted for, the target of 1,000PgC cumulative emissions begins to look even less certain to keep the increase in GMST below 2°C.

The likelihood that restricting cumulative emissions to 1,000PgC will keep the increase in GMST at or below 2°C is only part of the risk assessment.

We also need to consider how likely it is that we can actually restrict cumulative emissions to this amount. The examination of this task in Chapter 1 and in the next section, suggests that this is unlikely. If this unlikelihood is taken together with the 66 per cent chance that keeping within the 1,000PgC target will keep GMST rise below the target 2°C, we must conclude that the chances of averting dangerous climate change are low, possibly very low. That being the case, the logic for examining the possible benefits from some form of geoengineering in the policy mix becomes compelling.

The numbers

By 2012 some 380PgC had been emitted from fossil fuels and cement production, and a further 170PgC had been emitted from land use changes and changes in land cover. That leaves 450PgC to be emitted to remain within the 1,000PgC limit. In this reckoning, non-CO_2 emissions of a further $3PgC_{eq}yr^{-1}$ are ignored[2] because their forcing is largely offset by cooling aerosols.[3] If the current trajectory of CO_2 emissions continues unabated, the 1,000PgC will be reached well before mid-century.[4] It is impossible accurately to assess the likelihood that the 1,000PgC total will not be breached because it depends almost entirely on the rate of future uptake of low carbon technologies and this rests firmly in the realms of the uncertain and the surprising. Any estimates of this are in the nature of conjecture. What is known is that historically, new power sources have taken several decades to become firmly established. Figure 9.1 illustrates almost 240 years of US fuel consumption, transitioning from wood to coal and on through oil, gas and

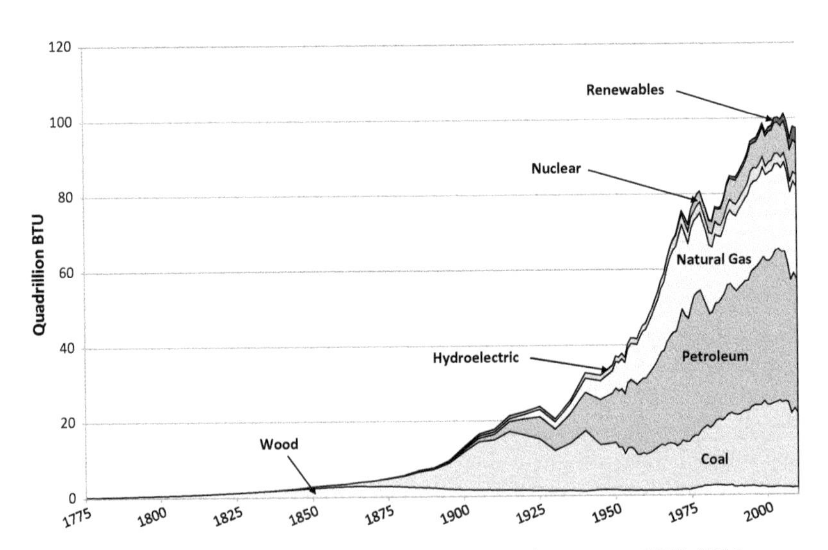

Figure 9.1 US primary energy consumption estimates by source 1775–2011
Source: US Energy Information Administration.[5]

nuclear towards renewables. There seems little reason to suppose that the uptake of renewable technologies will progress much faster than was the case for earlier new energy sources, but if it is assumed that they take only 40 years that would bring us to mid-century by which time, all other things being equal, the 1,000PgC limit would have been breached. However, the historical record greatly understates the renewables challenge. In the past, as coal gave way to oil and then nuclear, the old sources of power were not replaced but added to.

Figure 9.1 shows that in the US, all forms of fossil fuel used today are consumed at or close to historical peaks. Data for the world for the last half century show a similar pattern.[6] To achieve the required close-to-zero carbon future and stay within the 1,000PgC limit, renewables would need almost entirely to replace fossil fuels by the end of the century. This would represent an unprecedented rate of new technology penetration. Crucially, both charts show the magnitude of the task if the uppermost sliver for renewables is to grow largely to replace all the others except nuclear and hydro within the next 85 years. If the anti-nuclear lobby prevails, the challenge for renewables is that much greater.

The rate at which the take up of new energy sources takes place is not only a matter of research and development but is as much determined by the time it takes for the construction of the necessary production infrastructure at global scale and the penetration of the new products into the market place. The most rapid take up occurs where existing capital assets can be used to produce the new products and the new products can be used with little or no modification to existing end user equipment (machinery, vehicles, heating and so on). For renewables dedicated to replacing fossil fuels in central electricity generating stations, the burden of adaptation is likely to be modest. These have already accommodated shifts from coal, to oil, to gas and more recently to biofuels. These changes are transparent as far as the end user of the electricity is concerned but, as the biochar example in Chapter 2 and the discussion in Chapter 7 of CDR as a dominant response to global warming indicate, the availability of non-fossil fuel feedstock in sufficient quantity to displace fossil fuels is far from being secure.

The same technical ease of transition does not apply to changes in the power sources delivered to end users other than in the form of centrally produced electricity. For example, the shift from liquid hydrocarbon fuels to electricity, solar or hydrogen for powered vehicles or aircraft requires major changes not only to the installed fleet of vehicles but also to the systems by which the fuels are distributed. Electric cars are now beginning to be seen on our roads and it is likely that they will see dramatic growth in coming years. Yet until the electricity that powers them is also produced from zero-carbon sources, the transition will not be complete. These changes do happen, but historically, as the graphs show, they happen over decades. A major determining factor in this transition is the amortisation of the vast installed capital asset base that would become redundant as a consequence of this shift. The economic imperatives for not scrapping assets before the end of their useful working lives, that in the case of major infrastructure is usually measured in decades, is considerable.

These numbers do not prove that breaching the 1,000PgC threshold is inevitable, but they do show that it will require not only extraordinary political will and commitment but also resilience to uncontrollable exogenous factors – we must plan on the basis that the future holds some surprises and that some of them will not be pleasant. Dealing with climate change is not the only priority that politicians have, and for most it is not the one that keeps them awake at night.[7] It would seem prudent to marshall all the forces at our disposal and if geoengineering has the potential to contribute, one might suppose that our distant future successors would not thank us if we had failed to realise that potential in circumstances where there must have been reasonable doubt that emissions abatement at sufficient scale would proceed quickly enough.

There is an infinite number of emissions reductions trajectories by which we could cap future cumulative emissions to 450PgC, thus bringing the total to no more than 1,000PgC. Complex models are not necessary to understand that the longer the delay before we begin, in earnest, reducing CO_2 emissions, the greater the abatement burden we are bequeathing to the next generation and the more difficult and expensive it will be for them to achieve. Chapter 6 of IPCC AR5 WGIII, entitled *Assessing transformation pathways*, comprises almost 200 pages of detailed analysis, mostly driven by predictions from 'over 1000 new scenarios published since AR4'. The chapter seeks to answer this question:

> Is it possible to bring climate change under control given where we are and what options are available to us? What are the implications of delaying mitigation or limits on technology options?
>
> (IPCC 2014a, FAQ 6.1)

Its answer is that it is possible but only if we are able to achieve:

> large-scale transformations in human societies, from the way that we produce and consume energy to how we use the land surface, that are inconsistent with both [current] long-term and short-term trends.

It is unable to point to any of the many technologies for reducing energy demand or reducing carbon intensity that could currently be considered sufficiently close to deployment at sufficient scale to achieve any of the climate targets. Moreover, it notes that the scenarios that lead to atmospheric CO_2 concentrations of less than 450ppmv by 2100 all include some CDR to deliver so-called negative emissions to compensate for the expectation that actual emissions will not be reduced rapidly or deeply enough. Yet it also notes that there are considerable uncertainties about the timely and effective technical development of these CDR technologies at the required scale. The analysis in Chapter 1 explains that a zero-carbon target implies CDR of some $1.5PgCyr^{-1}$ if we assume that we cannot stop the emissions from land use and land use change. The computation in the Appendix to Chapter 2 illustrates the scale of that challenge.

With this context, Gardiner's caveat that the Arm the Future Argument is vulnerable to emissions abatement and adaptation being insufficient to avert dangerous climate change begins to look prescient. He suggested that if dangerous climate change were more imminent than most then thought, that geoengineering might be justified. As the previous few paragraphs make clear, more recent research does not suggest that the catastrophe is necessarily more imminent but rather that, whenever it happens, it is less likely to be avoidable by a reliance only on emissions abatement and adaptation.

While geoengineering would not be, and could never be 'our only option', it may be a vital part of a portfolio of responses in reducing overall risk and impact. Based on current understanding of the science, Gardiner *et al.* would be obliged to withdraw their objection to the Arming the Future Argument by invoking their own mixed portfolio caveat. The corollary is that the present generation should be expanding the options for future generations, not restricting them.

The emergence of systems thinking

A close reading of Chapter 6 in IPCC AR5 provides a salutary lesson in how far we are from having at hand the tools necessary to respond in a timely and effective manner to the threats of global warming. However, it proposes a policy approach that might have been taken from a systems thinking handbook. It says:

> Stabilizing atmospheric concentrations of GHGs and radiative forcing is a long-term endeavour. Whether a particular long-term mitigation goal will be met, and what the costs and other implications will be of meeting it, will depend on decisions to be made and uncertainties to be resolved over many decades in the future. For this reason, transformation pathways to long-term climate goals are best understood as a process of sequential decision making and learning. The most relevant decisions are those that must be made in the near term with the understanding that new information and opportunities for strategic adjustments will arrive often in the future, but largely beyond the reach of those making decisions today. An important question for decision makers today is therefore how near-term decisions will influence choices available to future decision makers. Some decisions may maintain a range of future options, while others may constrain the future set of options for meeting long-term climate goals.
>
> (IPCC 2014a, sec. 6.4.1)

The chapter also provides an unwitting illustration of the limits of reductionism. Much of the chapter is taken up with conventional predictive analysis, the import of which is to identify the scale of the challenges we face. But when it comes to policy formulation, the chapter recognises the need for 'a process of sequential decision making and learning', acknowledging that we cannot imagine a safe and happy future into being but must create it step by step, learning from our successes and failures as we go, in a heuristic process in which we must respect future generations and not usurp their agency by unduly constraining their options.

Conclusion

Paradigm shifts cannot be engineered, they just happen with the emergence of a critical mass of those that accept the new ways of understanding the world (Kuhn 1962). So it will be for the climate policymakers shifting from a linear reductionist to a heuristic systems approach. There can be little doubt that humanity has the ability to make this shift but I question whether our social evolution has yet progressed to a stage where we have developed the capabilities to do so. From our pre-human past we have developed innate and learned skills of social interaction to enhance our capacity to deliver our individual and collective needs. As our spatial reach has grown from the confines of small nomadic communities to the present globalised urban social world, we have been adept at developing institutions to govern the relations between individuals and between communities at all scales. The head of the family, the village elder, parish, municipal, national and international councils and parliaments, corporate boards, trustees, courts, UN agencies are some of those institutions. We have been less effective in addressing intergenerational issues. Edith Brown Weiss (1989; 2007) was amongst the first to articulate in depth the intergenerational dimensions of climate change. She proposed that a formal representative be appointed on behalf of future generations to attend to their interests in all international negotiations that might affect them. Jonas raised similar concerns noting that 'The nonexistent has no lobby, and the unborn are powerless' (Jonas 1973). Other institutional innovations could also be devised to recognise and neutralise the conflict of interest inherent in any present generation making policy that has significant transgenerational distributive consequences. That this has yet to happen illustrates the relatively underdeveloped state of our temporal relations with those in the distant future.

Never before has an issue like geoengineering arisen where the primary beneficiaries of our decisions are people in the distant future, entirely disconnected, except possibly in the broadest moral terms, from those making the decisions. Even emissions abatement and adaptation decisions could be understood, at least in part, to benefit some elements of the present human generation and other biota, but geoengineering under any realistic policy scenario is unlikely to be deployed at climate changing scale for several decades and would then need to be deployed continuously for many decades thereafter. The inherent inertia in the climate system, coupled with that of researching and developing geoengineering technologies, is such that the lapsed time from the decision to embark on a serious programme of empirical research to its deployment at scale is multigenerational.

A collective appetite to begin substantive empirical geoengineering research is not inevitable but will depend upon a complex mix of factors that might or might not come together. Whether one views this conjuncture as auspicious or inauspicious will depend upon one's ideological stance on geoengineering. Stilgoe (2015) insightfully concludes his book *Experiment Earth* by observing that 'The shared space of geoengineering research is an ideal location in which to rethink the relationship between science, politics and the public'. Geoengineering planetary processes is a literally extra-ordinary ambition that engages these three realms across a range of deep and timeless questions about our relations with the planet, with each other, with humanity's past and its future. But this should not be taken as an injunction to await the outcome of that rethinking before we act. Stilgoe captures the essence in his title; the Earth is an experiment.[8] The acting and rethinking go hand in hand as we edge our way clumsily, inefficiently, recursively, reflexively, and plurally into the *terra incognita* that is the future. Systems thinking tells us that the linear notion that we can first imagine the future we desire and then act so as to create it, is not only illusory, but also perhaps the source of the greatest risks we face.

Notes

1 '[T]he potentialities of nuclear-powered flight are so great that its continued development…is mandatory in the interest of national defense', stated the final report of the Nuclear Energy for the Propulsion of Aircraft program (cited by Stoffel (2000)). A detailed report in 1963 by an engineer in the Flight Propulsion Laboratory Department of General Electric examines the differences between chemical and nuclear powered rocket and aircraft engines. It concludes that nuclear reactors offer many advantages over conventional systems for certain space missions, available online at http://ntrs.nasa.gov/archive/nasa/casi.ntrs.nasa.gov/19640019868.pdf (accessed 24 March 2015). The aviation proposal was axed by Kennedy in 1961 (http://aviationweek.com/blog/1958-false-starts-aviation-s-atomic-age, accessed 24 March 2015). The nuclear rocket programme ended in 1973 (http://en.wikipedia.org/wiki/Nuclear_thermal_rocket, accessed 24 March 2015).

2 Data from US EPA, available online at www.epa.gov/climatechange/EPAactivities/economics/nonco2projections.html/ (accessed 23 September 2015).

3 One of the consequences of a major shift away from fossil fuels will be a significant reduction in atmospheric pollution particularly of sulphate aerosols that would have the short term effect of increasing global warming as these aerosols have a cooling effect (IPCC 2014b, FAQ 8.2, Figure 1; Matthews, Solomon, and Pierrehumbert 2012). It follows that the assumption that the non-CO_2 GHGs can be disregarded because of the effect of aerosols becomes increasingly less valid as the use of fossil fuels declines although Matthews *et al.* calculate that the effect will not be significant at least during this century. Nevertheless, this trend would place a greater burden on other policies to reduce GHG emissions or use CDR to

increase negative emissions to compensate (IPCC 2014b, Technical Summary Box TFE.8; Unger 2012).

4 Refer to Figures 1.1 and 1.2 in Chapter 1.

5 Data available online at www.eia.gov/totalenergy/data/annual/perspectives.cfm (accessed 23 March 2015).

6 Refer to Figure 1.5 in Chapter 1.

7 In the 2015 UK General Election campaign that is running as I write, there has been almost no mention of climate change as a key election issue. This is not because the politicians do not regard it as being of vital importance to future generations, but rather that they recognise that in the here and now of this election, it is not an issue that generates much interest within the electorate for whom the economy, the health service and immigration appear to be the concerns of the day.

8 When NERC (2010) used this title for a study of public perception of geoengineering, they ended it with a question mark as if to ask whether the Earth should be treated like an experimental laboratory. Stilgoe appears to have moved on and accepts that it is.

References

Brown Weiss, Edith. 1989. *In Fairness to Future Generations: International Law, Common Patrimony, and Intergenerational Equity*. Hotei Publishing.

Brown Weiss, Edith. 2007. 'Climate Change, Intergenerational Equity, and International Law'. *Vermont Journal of Environmental Law* 9: 615.

Charlesworth, Mark, and Chukwumerije Okereke. 2010. 'Policy Responses to Rapid Climate Change: An Epistemological Critique of Dominant Approaches'. *Global Environmental Change* 20(1): 121–129. doi:10.1016/j.gloenvcha.2009.09.001.

Gardiner, Stephen, Simon Caney, Dale Jamieson, and Henry Shue. 2010. *Climate Ethics: Essential Readings*. OUP USA.

Hulme, Mike. 2014. *Can Science Fix Climate Change?: A Case Against Climate Engineering*. First edition. Polity Press.

IPCC. 2014a. 'Climate Change 2014 Mitigation of Climate Change Working Group III Contribution to the Fifth Assessment Report of the Intergovernmental Panel on Climate Change'. Cambridge University Press.

IPCC. 2014b. 'Climate Change 2013 – The Physical Science Basis: Working Group I Contribution to the Fifth Assessment Report of the Intergovernmental Panel on Climate Change'. Cambridge University Press.

Jonas, Hans. 1973. 'Technology and Responsibility: Reflections on the New Tasks of Ethics'. *Social Research* 40(1). http://search.proquest.com.libezproxy.open.ac.uk/docview/1297204664/citation/646F320A878E4D58PQ/1?accountid=14697.

Kahneman, D., P. Slovic, and A. Tversky. 1982. *Judgment under Uncertainty: Heuristics and Biases*. Cambridge University Press.

Keith, David W. 2013. *A Case for Climate Engineering*. MIT Press.

Kuhn, T. S. 1962. *The Structure of Scientific Revolutions*. University of Chicago Press.

Lewis, Simon L., and Mark A. Maslin. 2015. 'Defining the Anthropocene'. *Nature* 519(7542): 171–180. doi:10.1038/nature14258.

Matthews, H. Damon, Susan Solomon, and Raymond Pierrehumbert. 2012. 'Cumulative Carbon as a Policy Framework for Achieving Climate Stabilization'. *Philosophical Transactions of the Royal Society A: Mathematical, Physical and Engineering Sciences* 370(1974): 4365–4379. doi:10.1098/rsta.2012.0064.

NERC. 2010. '"Experiment Earth?" Public Have Their Say on Technologies to Reduce Global Warming'. September 9. www.nerc.ac.uk/press/releases/2010/35-experiment.asp.

Stilgoe, Jack. 2015. *Experiment Earth: Responsible Innovation in Geoengineering*. Routledge.

Stoffel, Jesse. 2000. 'Dreams of Nuclear Flight'. University of Wisconsin-Madison.

Thompson, Michael, Richard Ellis, and Aaron Wildavsky. 1990. *Cultural Theory*. Westview Press.

Unger, Nadine. 2012. 'Global Climate Forcing by Criteria Air Pollutants'. *Annual Review of Environment and Resources* 37(1): 1–24. doi:10.1146/annurev-environ-082310-100824.

Epilogue

In October 2015 I was invited to participate in a three-day international workshop at Oxford University discussing options for greenhouse gas removal (GGR) technologies. The workshop brought together more than eighty people including many of the world's leading academics in this area, together with representatives from industry, media, UK research councils, the UK Met Office and philanthropic organisations, all seeking to promote GGR as a response to global warming. The discussions were fascinating and left me in a confused state of hope and despair.

Let me deal with the despair first. It was tragically reassuring to hear so many others echo my observations in this book that there seems to be a comprehensive global policy vacuum in regard to all forms of geoengineering, and in particular to CDR/GGR despite IPCC AR5 making it abundantly clear that staying below a 2°C increase in GMST demands a substantial contribution from carbon sequestration. Despite there having been a number of large-scale pilot projects for CCS, none of these is scheduled to be continued much beyond 2020. There is currently no significant public investment anywhere in the world in CCS or any other CDR/GGR technology. It will be interesting to discover what magic the world's leaders invoke at COP21 to reconcile their rhetoric with the science and technology.

Second, it was something of a revelation to discover that the extraordinary exploitation of shale gas in the US would not have happened without substantial investment by the US government in the 1980s and 1990s in early stage key enabling research, that would not have been financed by the free market because at that stage it was perceived as too risky (National Research Council et al. 2001; Wang and Krupnick 2013). This example illustrates the power that policymakers have to provoke the private sector into adopting new technologies. Conversely, the floundering European Emissions Trading Scheme (Borghesi and Montini 2015) underscores the point I made at the beginning of this book, that relying, as Margaret Thatcher did, 'on industry to show the inventiveness that is crucial to finding solutions to our environmental problems', is imprudent. Whether for shale gas or climate change, the short-term narrow commercial goals that characterise all private sector companies cannot be expected to deliver long-term public policy objectives without external priming to spark the necessary inventiveness into action. Until the international policymaking community recognises its indispensable midwifery role in relation to CDR/GGR technologies, it seems most unlikely that any serious progress will be made, a failure that will quite simply render unachievable its global climate goals.

On the plus side, I was truly impressed by the inventiveness of the academics in devising novel approaches to the twin challenges of sequestering atmospheric CO_2 and turning some of it into a raw material for products marketable at a scale sufficient to be climatically significant. There have been considerable advances in incorporating CO_2 into building materials and the manufacture of artificial liquid hydrocarbon fuels remains a priority for some. Moreover, developments in the technologies necessary to transport large volumes of liquefied CO_2 over long distances from the site of their production (whether from DAC, BECCS or similar front end processes) to the site of their sequestration in saline aquifers, suggest that this may be less of a problem than I have perhaps implied in this book. Perhaps my favourite technology, new to me but, like so many other 'novel' ideas, based on ideas that have been around for decades, was the solar chimney (Krätzig 2013). Solar chimneys use solar heating to create a powerful airflow through a very high chimney to provide a power source that drives electricity-generating turbines and in the process allows the air to be scrubbed of its CO_2 and other noxious and greenhouse gases. Some 70,000 of these could entirely replace fossil fuel electricity generation and could scrub the entire atmosphere every four or five years. Large as this number may be, it compares favourably with the more than 200,000 electricity-generating units currently installed worldwide.[1]

The three realms of academia, business and politics, typified by their respective drivers – knowledge, profit and power – need to be brought into closer alignment. I came away from those three days with immense optimism that we have the intellectual capacity to avert the worst ravages of global warming and a willingness of the academic and corporate realms to engage, but I also came away with a sense that politicians have yet to find a way of reconciling their short-term focus on retaining power with the implications of cleaning up the accumulated atmospheric CO_2 pollution. In most of the world we no longer dump our sewage in the streets or industrial chemicals in our rivers. Just because CO_2 isn't brown and smelly doesn't make it any less harmful. As I have commented in a footnote in this book, I do not mean to be critical of politicians; their policy paralysis is structural rather than due to ignorance, denial or incompetence. At this conference there was little expectation that COP21 would produce a globally binding agreement that even remotely makes the 2°C target likely; the general talk was of a 3°C to 6°C temperature increase. At the time of writing, we have only a few weeks to wait until we discover whether this pessimism is justified. For the politicians to engage globally in a meaningful way with academia and business to realise the potential we undoubtedly have within us to avert dangerous anthropogenic interference with the climate, requires systemic changes in their relations with their publics and the media. The inertia emanating from those with entrenched vested interests in the status quo makes those changes a slow and distant prospect, one that I fear is unlikely to come to pass soon enough to avoid a great deal of avoidable pain and suffering for proximate future generations. But that is no reason not to keep trying.

Note

1 Data from Platts, available online at http://www.platts.com/products/world-electric-power-plants-database, accessed 8 October 2015.

References

Borghesi, Simone, and Massimiliano Montini. 2015. 'The Allocation of Carbon Emission Permits; Theoretical Aspects and Practical Problems in the EU ETS'. Financialisation, Economy, Society & Sustainable Development (FESSUD) Project.

National Research Council, Committee on Benefits of DOE R&D on Energy Efficiency and Fossil Energy, Board on Energy and Environmental Systems, Commission on Engineering and Technical Systems, Division on Engineering and Physical Sciences, and National Academy of Sciences. 2001. *Energy Research at DOE: Was It Worth It? Energy Efficiency and Fossil Energy Research 1978 to 2000*. National Academies Press.

Krätzig, Wilfried B. 2013. 'Physics, Computer Simulation and Optimization of Thermo-Fluidmechanical Processes of Solar Updraft Power Plants'. *Solar Energy* 98: 2–11.

Wang, Zhongmin, and Alan Krupnick. 2013. 'A Retrospective Review of Shale Gas Development in the United States: What Led to the Boom?' *Resources for the Future DP*, 13–12.

Index